America's Round-Engine Airliners

Craig Kodera &
William Pearce

specialtypress
PUBLISHERS AND WHOLESALERS

Specialty Press
838 Lake Street South
Forest Lake, MN 55025
Phone: 651-277-1400 or 800-895-4585
Fax: 651-277-1203
www.specialtypress.com

© 2019 by Craig Kodera and William Pearce

Edit by Mike Machat
Layout by Connie DeFlorin

ISBN 978-1-58007-257-1
Item No. SP257

Library of Congress Cataloging-in-Publication Data
Names: Kodera, Craig, author. | Pearce, William, 1977- author.
Title: America's round-engine airliners : air frames and powerplants in the golden age of aviation / Craig Kodera and William Pearce.
Description: Forest Lake, MN : Specialty Press, [2019] | Includes bibliographical references and index.
Identifiers: LCCN 2018041355 | ISBN 9781580072571
Subjects: LCSH: Transport planes--United States--History. | Aeronautics, Commercial--United States--History. | Airlines--United States--History.
Classification: LCC TL685.7 .K63 2019 | DDC 629.133/343097309041--dc23
LC record available at https://lccn.loc.gov/2018041355

Written, edited, and designed in the U.S.A.
Printed in China
10 9 8 7 6 5 4 3 2 1

Specialty Press books may be purchased at a discounted rate in bulk for resale, events, corporate gifts, or educational purposes. Special editions may also be created to specification. For details, contact Special Sales at 838 Lake Street S., Forest Lake, MN 55025 or by email at sales@cartechbooks.com.

Front Cover: *Wright R-3350 radials powered the sleek Lockheed Super G Constellation, seen here in Trans World Airlines markings over lower Manhattan.*

Front Flap: *The Boeing 80A trimotor was a luxurious airliner for its time. Here, an example is loaded with baggage and mail before a transcontinental flight in the early 1930s.*

Front End Paper: *Douglas DC-7C "Seven Seas" was the world's first truly intercontinental airliner, able to span all major international air routes with ease.*

Back End Paper: *Perhaps the most identifiable piston-powered airliner ever built, Lockheed's triple-tail Constellation provided passengers with pressurized comfort.*

Title Page: *An actual color photo shows an American DST at Newark airport in 1937. The TWA rotunda can be seen in the background. (Photo Courtesy Boeing Co. via Jon Proctor with permission)*

Table of Contents: *A factory-fresh American Airlines Douglas DC-6 departs the manufacturer's home facility in Santa Monica, California, to begin its storied airline career.*

Back Cover
Left: *The Douglas DC-4 was a true sky giant in its day, as can be seen in this close-up of the former military Skymaster's engine nacelles and props.*

Top Right: *Lockheed 1049C Super Constellation in classic Trans World Airlines livery.*

Bottom Right: *Boeing's majestic Model 314 Clipper flying boat soars above scenic San Francisco Bay, inbound from another long-range journey across the Pacific.*

DISTRIBUTED BY:

UK and Europe
Crécy Publishing Ltd
1a Ringway Trading Estate
Shadowmoss Road
Manchester M22 5LH England
Tel: 44 161 499 0024
Fax : 44 161 499 0298
www.crecy.co.uk
enquiries@crecy.co.uk

Canada
Login Canada
300 Saulteaux Crescent
Winnipeg, MB R3J 3T2 Canada
Tel: 800 665 1148
Fax: 800 665 0103
www.lb.ca

TABLE OF CONTENTS

Acknowledgments 4

About the Authors 6

Foreword by Jon Proctor 7

Introduction 8

Chapter 1 | From Inline to Round 11

Chapter 2 | One Wing, Two Engines, All Metal 50

Chapter 3 | Presenting the Ship as an Airplane 78

Chapter 4 | Landplanes Become Viable Contenders 95

Chapter 5 | Expanding the Envelope 117

Chapter 6 | Some Serious Air Transports 136

Chapter 7 | Shrinking the Envelope 168

Chapter 8 | The Big Time 176

Chapter 9 | Twilight of the Goddesses 201

Epilogue | From Best of the Radials to the First Jets 211

Bibliography 213

Index 215

ACKNOWLEDGMENTS

CRAIG KODERA

Looking like one of those heroic images from the federal government during the Great Depression, this lovely study of man and machine is a TWA mechanic working on a DC-3 engine, perhaps at Kansas City, Missouri.

No one person ever writes or creates a book alone, and this volume is certainly no exception. It took the talents and institutional knowledge of several key men and women to bring this particular writing to light, and they need to be made known and congratulated.

First and foremost is editor and contributor Mike Machat. His concept for this book, along with publisher Dave Arnold, lit a fire underneath both authors and ignited a fabulous compilation of histories, facts, and interesting points of view. Mike was always there, always searching photo archives for me. His input on professional items was crucial to cobbling together all this raw material and making a cogent presentation. No book ever leaves the launching pad of Specialty Press without first getting the once-over by this editor. We have collaborated over many years and on many projects, and this one has been special.

Second, I must hold up to praise my coauthor, William Pearce. Bill, as I get to call him, is perhaps one of the most knowledgeable yet "readable" authors whose work you will ever have the pleasure of consuming. While I am pulling teeth to construct and illuminate each word on a page, Bill is just a house afire spewing a cornucopia of written language. And the frustrating aspect of his prowess for someone like me is that he makes it all look so easy! This book would not exist at all if it were not for Bill and his deep understanding of the complex and interwoven nature of the radial piston engine and its applications. He makes the extreme seem simple and easily understood. Thank goodness he agreed to sign on for this undertaking!

Next up, I want to thank several contributors. First among them is another old and dear friend and wordsmith, Jon Proctor. A former TWA cabin crewmember, Jon is the historian at large of the airline world and specifically his former airline. He too was an editor in a former life and knows his way around a page of type. He wrote both the foreword and a special article on the Lockheed Starliner. In addition, many of the photos contained herein are courtesy of Jon, regardless of the attribution appearing below them. His contributions have made this book utterly unique and as valuable as gold. I like to call Jon "my human encyclopedia of airline knowledge."

Test him sometime and you will be amazed by what he delivers. But never bet against him!

Adding to the sidebar content of the book is John Lewis, veteran pilot and all-around great human being. John is one of those few remaining air force aviators who experienced cockpit time in both recips and jets. His log book contains such airplanes as C-47s, B-47s, F-100 fighters, and Gulfstream business jets. He has been around the world and can tell you all about managing an airplane in various types of environments. John contributed a marvelous piece on the differences between the two DC-3 engine choices, Wright and Pratt, and why they mattered.

Speaking of photos, a most invaluable source has been tapped in Santa Monica, California, at the famous Museum of Flying located at the airport. The museum is the repository of most of the Donald Douglas and Douglas Aircraft archives and artifacts. This includes an astoundingly dense collection of images from over the decades. Thanks to Dan Ryan for all his help and work with my editor's quest for great photos. Also, a tip of the hat to aviation's old and constant friend Michael O'Leary, who owns and writes *Air Classics* magazine. If you think the National Archives has historical photos, the main office of Challenge Publications has a vast number of files containing priceless pictures of aviation's past. I am just thrilled that Michael saw his way clear to help in our project.

One of the most pleasurable aspects of this book has been tapping the life experiences of my lovely and loving friend, Judy James. Judy was there and did it: flying as a stewardess for Eastern Airlines during the 1950s, the Rickenbacker/Dick Merrill era. Wow! She is old school and here on these pages just for you. We had a blast recounting her time in the air, flying the crazy days and with the great old men and women of aviation lore. Man, oh man: to be alive in those heady times and a part of making history. Thank you, Judy.

Finally, there are what seems like zillions of individuals who allowed the use of their photos and photo collections. They all will receive a free book because of their generosity. Please notice the many names tacked to the precious images that absolutely make this book a great keepsake.

All these wonderful people and indeed this entire project were, when we truly take the time to consider the how and why of a creative effort such as this, ultimately the work and vision of our God who makes all things possible. I always bear this in mind and am forever grateful to Him.

These young aeronautical engineers of the future listen attentively to their instructor as he describes the typical aero engine and its functionality. The photo was taken in 1942.

WILLIAM PEARCE

This book represents the combined efforts of many. Seeing our ideas come together has been extremely rewarding, and I hope that those reading the work will enjoy it as much as I have enjoyed watching it take shape.

I'd like to thank Craig Kodera for being a fantastic partner as we brought this book to life, Mike Machat for having the patience to let Craig and me run wild and for his willingness to polish our work to a high luster, Kim McCutcheon for his eagerness to help in any way, and the Aircraft Engine Historical Society for providing documents, information, and images.

Special thanks go to my wife for her support, encouragement, assistance, and understanding.

ABOUT THE AUTHORS

The sign of dependable engines: Pratt & Whitney and their eagle. A classic moniker on one of the world's classic airframes, a Beechcraft D-18.

CRAIG KODERA

Born in Riverside, California, in 1956, Craig grew up in an aviation family in the heart of aerospace country: the Los Angeles basin. From his father, who worked in the aircraft engineering industry, to his uncles, one of whom was a Brigadier General in the air force and another involved in the operation of Los Angeles Airways helicopter airline, Craig became an aviation fanatic from his earliest age. The family home was a scant 2 miles from LAX, which meant plenty of "airport fence" time for Craig.

After attending UCLA and graduating with a BA in Mass Communications and Art History, Craig took his ROTC experience and attended Officer Training School in 1979 followed by air force pilot training at Reese AFB, Texas, as a member of class 80-08. His military flying career included the HC-130 Hercules for the 303rd Air Rescue Recovery Squadron and the KC-10 Extender for the 79th Air Refueling Squadron, both in the Air Force Reserve and both at March AFB, California.

In 1986, Craig was privileged to be hired at Air California (Air-Cal) and assigned to the 737-100/200/300 series of aircraft. In 1987,

American Airlines purchased AirCal and Craig eventually spent 5,000 hours flying the MD-80 airplane series for that company. Today, Craig is medically retired from flying but is still heavily involved with aviation and the people who make it happen.

Painting fine art since the age of 14, Craig has enjoyed a full second career as a freelance artist. Specializing in aviation art, Craig's client list includes McDonnell Douglas Helicopter, Airbus North America, Northrop Grumman Corporation, Bell Textron, Federal Express, Smithsonian National Air & Space Museum, San Diego Aerospace Museum, the National Museum of Naval Aviation, Pelican Publishers, Republic Press, and Specialty Press.

Craig is proud to have been the Charter Vice President of the American Society of Aviation Artists and an artist of The Society of Illustrators, Los Angeles. Craig is also a Life Loyal member of the Sigma Chi fraternity.

Awards have included Best of the Best from *Aviation Week* magazine in 2000, the prestigious R. G. Smith award for Excellence in Naval Aviation Art in 2001, and the Nixon Galloway Golden Age of Flight award in 2011.

Craig has been associated with the Greenwich Workshop since 1984 and his limited-edition prints and posters are found across the country at Greenwich dealer galleries.

In addition to painting, Craig has recently been involved with the illustration and layout of children's books and has also authored four adult volumes. All his work can be found on Amazon.

WILLIAM PEARCE

William's interest in aviation began at a very young age when he lived at a fly-in community with his parents. William is a mechanically minded individual whose primary interests are aircraft piston engines, World War II aircraft, and air racing. He is a graduate of Cal Poly San Luis Obispo, a lifetime member of the Aircraft Engine Historical Society, and a crew member for a record-setting air racer.

Over the years, William has amassed a large collection of aviation literature. He wrote and published *Duesenberg Aircraft Engines,* and he has published more than 150 articles on his website (oldmachinepress.com). William lives on California's central coast with his wife and children.

FOREWORD

Today we are able to fly nonstop for 18 hours or more across oceans and continents aboard modern twin-engine jetliners. Barely more than 30 years ago, the idea of crossing the Atlantic aboard a three-engine passenger jet was just beginning. But imagine what it was like in the late 1940s, traveling from the U.S. West Coast to Hawaii in a bit under 10 hours, carried by a piston-pounding DC-6. Earlier-generation DC-4s did it in half a day. Either way, it was a long ride.

Remarkable progress in air travel can be attributed to a common thread: the advancement of engine powerplants, from rudimentary noisemakers to ultra-reliable turbine engines capable of routinely transporting remarkable payloads around the globe at 10 miles a minute.

With this volume, authors Craig Kodera and William Pearce bring us from the commercial air transport's very beginnings to the zenith of what is fondly referred to as flying in the "Golden Age of Air Transport," when jetliners were still on the drawing board or in early production, and still a dream to come for passengers and crew.

In the time period of "iron pilots and wooden airplanes," engines were incapable of extended use, required constant maintenance, and could quite literally come apart, often in flight, and with alarming frequency. Then, gradually, aeronautical engineers mastered the challenge of providing not only in-flight reliability but also constant use over longer distances.

Once people began accepting air travel as more than a chancy proposition, aircraft reliability and economy began to provide the prospect of safe and profitable operations for airline pioneers who believed that the airline industry really did have a future.

One of the biggest steps forward came out of necessity when a thought-to-be reliable Fokker F-10A, operated by Transcontinental & Western Air (TWA), suffered a wing failure in March 1931 and plunged to the ground near Lamar, Kansas, killing all eight aboard. Among the dead was beloved Notre Dame football coach Knute Rockne. Although wood rot was blamed for the accident, Rockne's shocking death effectively sealed the fate of wood construction in civilian airliners. Fearing the very demise of his company, Jack Frye, TWA's visionary president, petitioned aircraft builders to produce all-metal designs, which led Douglas President Donald Douglas and his engineers to develop the Douglas Commercial One (DC-1), from which the DC-2 was born, followed by the revolutionary DC-3.

Along with this historic step forward in airliner design came new engines so reliable for their time that Douglas talked Frye into a twin-engine airliner instead of the requested concept with three powerplants. From then on, engines equaled or surpassed the need for sturdy airframes.

Fast forward to World War II. With its necessity of even stronger designs to carry heavy bombers great distances, you learn here how such advances carried over to the civilian market. Military C-54s, designed by Douglas and built under contract by other manufacturers as well, served as civilian DC-4s until newer, more powerful, and pressurized DC-6s came on line. Boeing's mighty B-29, extensively modified, emerged as the luxurious model 377 Stratocruiser. On the other hand, Lockheed's graceful Constellation became the C-69 military variant before it could be placed into civilian operations following the hostilities. After the war, it was put into passenger service and flown all the way to Europe, albeit with fuel stops.

In each of these examples, it was further development of aircraft engines that provided greater reliability and range along with profits for their builders and customers.

So come with the authors as they tell this amazing story of American ingenuity and determination as an industry advanced from varnish and bailing wire to that Golden Age of Air Transport, which for those of us who lived through it, will never be forgotten.

Jon Proctor
Sandpoint, Idaho

Who says the depression era was colorless? Here are a couple of beauties posing in 1941 for the Neiman Marcus department store in Dallas, Texas. The airplane is a Lockheed 10 Electra, which would put a smile on anyone's face.

INTRODUCTION

"Turning one!" The heart of this book's writing comes to life on the Breitling Super-G Constellation in Switzerland, as oily and smoky today as it was when new in 1956. (Photo Courtesy Aldo Bidini)

The happy face of our air hostess is welcoming indeed as we approach a Delta Air Lines DC-3 for our journey to days gone by.

CRAIG KODERA

Noise! Smoke! Flames; clatter; oil everywhere. What in the world *is* that contraption attached to the wing of that old airplane and why is it there? It is an internal-combustion radial-piston aero engine, otherwise affectionately referred to as a "round engine" by pilots and mechanics alike. This type of power-plant was the only real game in town from the early 1920s to the late 1950s, and if you wanted to move people and cargo around the planet and do so efficiently and expeditiously, this was the accepted solution. From our vantage point some 60 years later, with turbine power now the accepted norm, it seems almost ridiculous to have used radial engines, but as historians, we have to constantly remember this salient point: *It's all we had!*

Human history is a series of building-block achievements, each one acting as a foundation for the next as new improvements come to the fore, supplanting those that came just prior. "How did we ever get along without such-and-such invention?" is the question that immediately follows. Just like the answer to a brain-teaser riddle, the latest and greatest innovation seems so obvious in its elegant solutions that we cannot comprehend why we didn't simply cut out the middle man and just invent *it* in the first place. Of course, the process doesn't work like that.

What does take place is a slow and methodical march, agonized over by great minds of the times, and in the world of aviation, sometimes aided by the ultimate sacrifice of those who tested those fledgling concepts. The piston engine aircraft powerplant is the perfect example of this scientific and humanistic collaboration, with its end-game result being the safe and (for its time) reliable performance as a means to move passengers and merchandise swiftly to their destinations every day.

This book guides you along two parallel tracks simultaneously. It discusses the inner workings and development of piston engines over the decades with just enough technical information to keep the engineers and mechanics among you happy, and then looks at the practical applications of the technology and places you in the various aircraft and their operating environment to give you the full effect of flying in a piston-powered airliner. If this is your first time as a passenger in the golden age of air transportation, you are in for a real treat and some eye-opening surprises.

Plenty of fresh and sometimes unique/obscure photos are included in this book, some more than 100 years old, all of which paint a picture of the Great Journey through time that will now include you. From the shop floors at Pratt & Whitney and Curtiss to the vibrating, upholstered cabins of the world's great airliners of the 1920s, 1930s, 1940s and 1950s, get ready to experience the past and smell, feel, and know in your heart and mind what it was like to travel by air back then. Fasten your seat belt!

WILLIAM PEARCE

The Boeing 787 is the most advanced airliner currently in service. Yet, despite the use of composites, computerized avionics, and advanced flight controls, the 787 looks remarkably similar to a two-engine Boeing 707, an aircraft designed in the mid-1950s and America's first jet airliner. During the past 60 years, the general design of the airliner has been more about refinement than revolution.

By contrast, the Boeing Model 80 of 1929 bears little resemblance to the Boeing 377 of 1949. The 80 was a fixed-gear, trimotor biplane capable of carrying 12 passengers at 125 mph over a range of 460 miles. Just 20 years later, the 377, a pressurized, four-engine monoplane, was in service and capable of carrying 100 passengers

Fifty-two years separate the first flights of the Boeing 707 (left) and 787 (right). Although a myriad of advancements have been made, such as computerized flight controls and avionics, composite construction, and incredibly efficient engines, the two aircraft appear remarkably similar. (Photo Courtesy Boeing)

The Boeing 80 (left) first flew on 27 July 1928. The fabric-covered biplane was powered by three 9-cylinder 410-hp P&W Wasp engines. The Boeing 377 (right) made its first flight on 8 July 1947, powered by four 28-cylinder 3,500-hp P&W Wasp Major engines. A comparison of the two aircraft illustrates the amazing advancements in aviation that occurred in fewer than 20 years. (Photos Courtesy William Pearce Collection)

at 340 mph over a range of 4,200 miles. In those same 20 years, advancements in aviation enabled the 377 to carry 833 percent more passengers, 272 percent faster, and 913 percent farther than the 80. That level of progress represents far more than just the continued refinement of an aircraft; it was a downright technological revolution.

A number of factors combined to make possible the revolution in transport aircraft design spanning from the 1920s into the 1940s. Chief among these factors was the development of the radial aircraft engine. Improvements in the understanding of combustion, metallurgy, fuels, oils, and of engine design, construction, and cooling enabled a transition from unreliable, heavy, 400-hp engines to dependable, highly refined engines capable of more than 3,000 hp.

Powerful and reliable radial engines gave aircraft designers confidence to create larger and more capable aircraft. Each new generation of transport aircraft pushed the boundaries further, until nonstop flights between continents were possible. However, there were limits. Radial engines were heavy and complex, with hundreds of moving parts. Although these well-engineered miracles could run hundreds of hours without a fault, they did not always do so. Their limitations led to the jet-engine revolution, and radial engines became a mere footnote in most texts.

A number of books lightly cover the piston-powered airliner story, but *America's Round-Engine Airliners* supplements that story with a close examination of radial-engine development. A handful of pioneers laid the foundation for all radial-engine development, and each advancement was built on the knowledge gained with previous designs and was made possible by the continued refinement of machining and manufacturing techniques.

Over the years, both Wright Aeronautical and Pratt & Whitney Aircraft used a few different naming conventions for their engines. Even so, most people referred to an engine by is military designation of an "R" (for radial) followed by the engine's displacement in cubic inches. In this book, the military designations are used the most

frequently for clarity and simplicity. You may not know what an "S1C3G Twin Wasp" is, but using "R-1830" gives you some idea of the engine and how it may compare to other engines designated in the same manner.

A few individuals are mentioned in the following pages, but they are just a small representation of the thousands of skilled machinists, draftsmen, engineers, workers, and managers who helped build air travel. It would be an overwhelming task to mention all of the skilled men and women who had a hand in the development of the radial engine and would be just as daunting to convey all of their many setbacks and triumphs encountered along the way. It is my sincere hope that you will forgive any omissions and be delighted by many new discoveries.

Strap yourself in, the weather is fine and the crew is keen. Glad to have you with us for this book. Welcome aboard!

Chapter 1

FROM INLINE TO ROUND

A scant 15 years after the Wright Brothers' first flight in 1903, here is the shape of the quintessential fighter airplane of World War I: a Nieuport 11. (Photo Courtesy Andreas Zeitler)

This stearman 4DM-1 Senior Speed Mail is powered by a P&W R-985 Wasp Junior, 450 hp, one passenger. (Photo Courtesy Addison and Ryan Pemberton, owners, pilots, and photographers with permission)

Commercial air travel had a very discernable starting point more than a century ago. In the 1920s, an entire industry blossomed with an accompanying manufacturing infrastructure to accommodate its mercurial growth. The rapid advancement in aerodynamic and powerplant technology in this era was exponential from today's vantage point. As you read this book, you will be awestruck by what inspired visions led to in this time period.

Context is everything in history. For all the complaints about modern-day air travel, we must hew to this notion of relativism and bear in mind just how primitive commercial flying was in those first decades. From this perspective, we can see that today's airline experience is like Star Trek by comparison! Let's engage our larger survey of the past and examine airplanes of wood and wire, cloth and castor oil, open cockpits, and men of daring-do.

Getting There by Air

The 1920s

To set the stage on the 1920s, focus on the World War I conflict of the previous decade and the fact that man had also just started

This current-day poster has the period look of 1920s Southern California and the de Havilland airliner by Kerne Erickson. (Copyright Greg Young Publishing Inc.)

Looks like an open house with everyone coming to see the Douglas Cloudster of Claude Ryan's Los Angeles-San Diego Airline. This airplane is the first aircraft solely designed and manufactured by Donald Douglas and his Douglas Aircraft Company of Santa Monica, California. Ryan modified the airframe to accommodate 11 passengers in wood-paneled luxury, being truly ahead of the curve in air transport technology. (Photo Courtesy AAHS)

actively leaving the ground beginning in 1903, some scant 20-some years prior to organized airline companies. The idea of routinely taking an airplane from one point on the map to another while on a schedule seemed like pure folly, and a life-ending one at that!

Rather, considering the movement of paying passengers on their airplanes, the various few businesses involved in the nascent world of commercial aviation in the 1920s focused first and foremost on flying the mail. Indeed, the US Postal Service, not terribly old itself after the war, became quite visionary in 1918 and took upon itself the task of using war surplus airplanes such as DH-4s with all those Liberty engines on hand, for moving mail across the country.

From that starting point to early 1927, the Air Mail Service was instrumental in establishing the very foundation that enabled commercial air traffic to be born. Entire infrastructures had to be developed and built. Innovations included landing fields with permanent buildings connected by telegraph and telephone communications; weather forecasting and dissemination; refueling, of course; and most important of all, lighted airways. This last item moved aviation

The de Havilland (AIRCO) DH-4, a prolific airplane in the immediate post–World War I period, was the primary recipient of the Liberty engine. Lack of reliability made the biplane a less-than-sterling member of the fleet in the early days of airmail delivery in the United States. Of the five manufacturers of this engine, Packard was the best with the highest-rated quality-control checks during assembly. Had Packard been given a sole-source contract, perhaps the history of "America's Engine" would have been different. (Photo Courtesy AAHS, Charles Stewart)

Colonial, based in Buffalo, New York, won CAM number 20 and linked that state via air delivery in some of the country's worst weather. Look closely at the two pilots' helmets and note a plug-in for aviation's newest innovation: two-way radio communication! The Mailwing was powered by the familiar Wright J-5 Whirlwind.

from daytime-flight-only to round-the-clock schedules as airplanes traversed the wide-open expanses of the far west of the country where there were no cities and therefore no visual checkpoints to guide the aviators along their routes.

The lighted airway concept eventually spread to much of the remaining parts of the country and, until radio navigation aids were introduced, were often the sole means of visually plotting a course (along with landmarks and railroad lines!). By 1926, the USPS was beginning the move away from mail carriage, divesting it to private-enterprise commercial lines. Two other important events were now needed (and occurred), which made this change possible.

The US congress enacted in 1925 the Kelly Act, a piece of legislation that was the first major step toward creation of an airline industry. This bill was the honey pot that made the commercial business now go. It provided for the transfer of the mail service to private carriers, contracts to be awarded via a bid process. Following close on its heels was the Air Commerce Act of 1926, which instructed the Secretary of Commerce (at that time Herbert Hoover) to designate and establish airways, to organize air navigation, to arrange for research and development of such navaids, to license pilots and aircraft, and to investigate air accidents. This was the foundation for the government organ known for years as the Civil Aeronautics Authority (CAA), later changed to the now-familiar Federal Aviation Agency (FAA), still responsible for all the above activities. Now the United States had a mechanism to encourage, guide, and strengthen air commerce.

Stepping Up to the Plate

As the country became serious about hauling money-making cargo (the mail), several bright entrepreneurs came forward to stake their claims to the future. First on the list was the always innovative Henry Ford. He teamed up with William Stout to create the first in a series of all-metal airplanes in the United States, the Stout 2AT. The route they bid for and won spanned the distance between Detroit and Chicago and later between Detroit and Cleveland.

Nearly simultaneously, and out West, Walter Varney won the route from Boise, Idaho, all the way to the Seattle area. He used Curtiss Swallows. And this is where the new radial engine made an appearance in the form of the Wright J-4 Whirlwind.

Also, out West appeared Western Air Express, winning its route from Los Angeles to Las Vegas to Salt Lake City. Flying from the now-defunct Vail Field in Southern California, Western began carrying passengers shortly after inaugurating its route, and for all of 1926 transported 209 individuals for the astounding price (then) of $60. For the north-heading route from Los Angeles to Seattle, Pacific Air Transport, flying Ryan M-1s, carried a passenger for $132. The total trip time took 18.5 hours.

Down South, the small Florida Airways ran mail and passengers from Atlanta, through Jacksonville, Tampa, Fort Myers, and ending in Miami. The airline used the Stout (Ford) 2AT.

Then there was the Philadelphia Rapid Transit Service (PRT), which flew from the Philadelphia Navy Yard to Hoover Field in Washington, DC, later including Norfolk, Virginia. The company actually placed more emphasis on transporting passengers than mail. Flying the Fokker VIIa, a trimotor airliner from Dutch designer and well-known aeronaut Anthony Fokker, PRT flew 3,695 folks during its first five months and charged $15 one-way fares. The operation was technically successful, being run as any real airline we know today with thrice-daily schedules between points. Unfortunately, overall the business lost money.

PRT had truly demonstrated that traveling by airplane could be safe and reliable, but the perception by the public was still of a risky, chancy, stunt-like undertaking. As a passenger, you would be more often than not no different than the pilot, both in terms of

Here is the first all-metal airliner in the United States: the Stout 2-AT, designed by William Stout of Michigan. Stout's designs, including cars, captured Henry Ford's imagination and business sense, so he bought the company. Always the innovator, Ford bid on and received the first airmail contract from the Post Office Department and began service with this airplane, Maiden Dearborn. The metal design was no doubt influenced by the world's first all-metal airplane, the German Junkers F-13, also of corrugated construction.

clothing as well as danger. If this was airline service, well, who wouldn't rather walk! It seemed that flying just wouldn't catch on. That was, until May 1927.

Pivot Point

What did it take to ignite the imagination and the passions of the general public? How did the desire and need manifest themselves so that the entrepreneurial capabilities of some very far-seeing men could come to fruition? The answer is, in a word: Lindbergh. In the spring of 1927, the young pilot braved crossing the Atlantic Ocean by himself in a single-engine airplane, flying nonstop for nearly a day and a half. "New York to Paris!" was the latest conquest for aviation, and the world sat up and took notice. This was the providential opening for the business of passenger aviation to literally take off.

And a real business it was, now being formed and shaped by men of industry and transportation. In the Northeast United States the players included the famous Daniel Guggenheim, tycoon and philanthropist, who created a trust to spur commercial aviation. At the same time, the president of the Pennsylvania Railroad (PRR), William Atterbury, envisioned a cooperative venture between his railroad and passenger-carrying airplanes to span the entire American landmass. (We can assume that the railroad's prototypical experience with PRT laid the foundation for what was next to come.)

Joining Atterbury in his dream of a transcontinental rail-air service was Paul Henderson of National Air Transport (NAT), which ran an airmail route between Chicago and Dallas. Then Henderson's boss and owner of NAT, another tycoon and financial wizard, Clement Keys, managed to wangle Charles Lindbergh as technical advisor to the soon-to-be-named Transcontinental Air Transport (TAT). What a feather in the cap to have that particular name associated

The incomparable Charles "Lucky Lindy" Lindbergh in 1928, stands before his Curtiss Falcon biplane, which he flew on route, surveying and proving flights for his new consulting role with TAT as technical advisor. To be so accomplished and respected at the young age of 25 speaks volumes as to the impact this man had on aviation. A former airmail pilot, Lindbergh knew about the equipment needed to operate in an open-air cockpit in the dead of winter as testified to by his flying garb. (Photo Courtesy Smithsonian National Air and Space Museum CW8G-M-0470)

prominently with the new airline. Financed with PRR money to the tune of $3 million, the company was established in May 1928, just one year after the crossing of the Atlantic.

As envisioned, the rail service would cross the Allegheny Mountains on the route from Philadelphia to Columbus, Ohio. At that point, the airplane would cover the next leg of the journey from Ohio to Waynoka, Oklahoma. Then, you would take the train again

TAT Plane-Train Routing
Train Service in Blue

Pennsylvania Railroad
Santa Fe Railway

You can certainly tell who was promoting air-rail service in this poster, and it wasn't the airline! Stressing safety and comfort, marketing men at the railroad knew their potential customers still considered flying to be a dangerous undertaking and played to that fear. The blue areas represented the overnight rail service portions of the trip.

Whether they knew it or not, the railroad men of PRR were actually building an airline system from New York to Los Angeles and beginning the process of putting themselves out of business. Again, the blue sections of the route were by train, which were supplanted by round-the-clock air service as aids to night flying became reliable and plentiful. Note the near proximity to the Western Air Express (WAE, later Western Airlines) route system starting at Kansas City and going west. This overlap was not lost on the central planners of the Post Office Department in Washington, DC. See page 19. (Map by Craig Kodera)

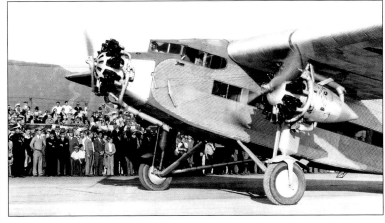

An eager crowd at Glendale stands by to see Lucky Lindy pilot the first flight of TAT's coast-to-coast service. In those days, crowd control was not nearly what it is today. (Photo Courtesy Museum of Flying Collection)

Charles Lindbergh is surrounded by TAT and PRR execs prior to first route-proving flights in the Ford. Best guess is William Atterbury, PRR president, standing just to Lindy's right. It's easy to see why the famed aviator was nicknamed "Slim." Note the corrugated metal construction of the Ford Trimotor (Stout). (Photo Courtesy Smithsonian National Air and Space Museum 92-3117)

Much like the WAE foldout brochure on page 19, this TAT example is just as informative to the prospective passenger. Comfort and safety are mentioned twice, modern aviation technology is highlighted, and the Lindbergh name is dropped conspicuously within the text to remind folks of why they should feel confident in flying with TAT. Even as early as the 1920s, Americans were in a hurry, and the air-rail service was there to speed the movement of "busy people" from coast to coast.

The fuel tank for the center engine was right behind the firewall in the nose. This particular trimotor has the often-seen early forward-swept windscreen design prevalent on early airliners.

(Santa Fe Railroad) to Clovis, New Mexico, and an airplane to Los Angeles (Grand Central Air Terminal, Glendale). Service began with Lindbergh at the eastbound airplane's controls on 7 July 1929, and was scheduled to last 48 hours to complete the entire cross-country journey, the rail portion taking place at night with customers in sleeping cars. The airline did not want to risk passengers in the treacherous hours of darkness for airplanes.

TAT ran a very successful and novel operation until 25 October 1930, when the train portion of the route was dropped. Trimotors now flew the entire length of the trip, with an overnight at Kansas City, Missouri. By 1932, even that stop was removed, and it was all airplane, all day and all night (thanks to lighted airways and radio navigation instruments). The trip now took less than 36 hours to complete but was no doubt arduous in the "Tin Goose" with its lack of heat and noise muffling. Life in the air, it seems, was only for the stout at heart.

Stepping backward now to the first days after Lindbergh, activity on the West Coast was heating up. The Guggenheim Trust was presented to an airline operation being run by Harris Hanshue, who operated Western Air Express on the aforementioned Los Angeles to Salt Lake City mail route. The intent of the award was to create a "Model

Both passenger airlines on either coast provided superb, white-glove service to their passengers, and it started with limousine service from city center points (hotels and ticket offices) to the airfields. TAT used a standard coupe pulling a wood and metal trailer that accommodated the outbound flight's fares. Known as the "Aero Car," examples of this confabulation were in place at every TAT stop across the nation. (Photo Courtesy AAHS)

The start of the air leg of the route was Columbus, Ohio. Note the rail spur running right to the front door of the terminal building. Passengers "detrained" and walked under a covered portico to reach the airplane, visible to the left of the terminal. This terminal exists today in restored beauty on the east side of the current airport. This is the westbound first flight on 8 July 1929, and as was the case in California, the crowd is milling about while engines are running. (Photo Courtesy AAHS)

Airway" that specialized in flying passengers, catering to them with top-notch inflight and ground service. True airliners were purchased from Atlantic Aviation, which built Fokker airplanes in the United States. The F-10 trimotor fit the bill, and passengers were treated to box lunches catered by the Pig 'n Whistle restaurant in Los Angeles and served by the first flight attendants in the country, male stewards. Fares were also chauffeured by Cadillac limousine to the airports. Western's model route was from Los Angeles to San Francisco with a stop at Fresno.

As it looked in 1929, this Ford 5AT Trimotor restoration of the "City of Columbus" is completely representative of the fleet used at TAT for its transcontinental route system. (Photo Courtesy AAHS)

Flying was exciting in the roaring 1920s. These two young ladies have exuberant expectation and perhaps a little terror written all over their faces. The Colonial Airlines Ford Trimotor is heading from New York to Boston for a weekend frolic in the Bay City. In aviation's early days, an airport was referred to as a "landing field" or "airfield." Judging by the surface beneath the ladies' feet, you can guess why such names were appropriate. Note the wicker seats.

Is that all there is to the Ford cockpit? Yes, with the exception of engine instruments located on the landing gear/engine strut just outside the pilots' windows. Automotive and marine industrial design elements can be seen in the layout of this control cabin.

In all its turn-of-the-century luxury, this restored Ford interior provides an understanding of the appointments and ambiance of flying in a TAT trimotor. Note the wood paneling and individual sconce light fixtures above the seats. This configuration is actually a bit later in the airplane's life as the seats are metal buckets with leather backs and seat cushions. Original delivery outfitting used wicker seats! (Copyright by David Schultz)

This poor-quality vintage photo shows in-flight mealtime aboard a TAT airliner: finger sandwiches, a relish side, a fruit cup, and a glass of milk. Note that all the place setting items were china, silver, and glass laid on linen tablecloths. Heaven knows where the steward kept all this in such a tiny airplane! Look too at the inlaid wood wall paneling, sconce fixtures, and curtains (also shown in the restoration). Noise levels were nearly deafening, however.

Pioneering at Western Air Express

Both TAT and WAE became busy acquiring other airline operations and additional mail routes in those first two years. By the end of 1930, in order to not compete on the same single airmail route, Postmaster General Walter Folger Brown insisted that they merge in what was coined the "shotgun marriage" of airlines. The combined airline, which spanned the continental United States from San Francisco and Los Angeles in the West to New York and Philadelphia in the East was now, indeed, the model airway of its time.

A quick survey of some of the goings-on from airlines around the country and the aviation industry in general yields several airplanes and airlines. Some were niche, to be sure, but nevertheless quite significant, such as Colonial Airlines and Pacific Air transport/United Air Lines. These fledgling industries were experimenting with many answers to the propositional question of how to fly passengers and mail at a profit.

This original foldout page from the 1929 Western Air Express Schedule and information brochure depicts WAE's original role as Guggenheim's envisioned "Model Air Line" flying up the California Corridor (as known in later years). That route stretched from Vail Field in Alhambra to the Oakland Municipal Airport in the eastern Bay Area. The equipment at that time was the 14-passenger Fokker F-10 trimotor, as seen here. The inset shows advanced features of the airline. Artist Ruth Taylor White truly captured the essence of 1920s California with her illustrations of people and locales.

Early in 1928, Western also had purchased Pacific Marine Airways whose sole mission was to fly from the Los Angeles basin to the island of Santa Catalina, 26 miles off the coast of Southern California. Besides Loening amphibians, Western used a single Sikorsky S-38 to shuttle vacationers to Avalon Harbor and later to Pebbly Beach near the Hotel Saint Catherine resort. This was possibly a staged photo op for Pennzoil products.

The airlines' bread and butter for many years was hauling the US mail. Sacks and parcels are being loaded into one of the former WAE Fokker 10s. When this photo was snapped, the forced merger between TAT and WAE had taken place, with the new entity known as Transcontinental and Western Air Lines (TWA). Notice all of the component air carriers receiving billing under the consolidated name. Note too the way Fokker mounted an entire wing simply atop the boxlike fuselage. This view shows the tower and terminal at Alhambra, Western's main base of operations.

Not to be outdone, Boeing's Hamilton division produced the H-47 Flamingo transport, also of corrugated aluminum skin. The largest operator of the type was Northwest Airlines, which used the airplane starting in 1928 along the mail route from Chicago to Minneapolis. This airplane is at a typical airport along that route system, dealing with typical weather as well. That's a snow blower at the tail of the airplane. The Metalplane, as it was known, was powered by a single Hornet engine of 525 hp and carried six passengers. Interestingly, the designer of this airplane was James McDonnell who went on to form his own company that later merged to create the giant McDonnell Douglas Corporation. (Photo Courtesy AAHS)

Airmail was a definite business opportunity for the various carriers around the country. Here is a later look at a portion of the Colonial (CAM 20) fleet. Present are two Sikorsky S-38 amphibians along with a Fairchild FC-2. The engines represented here are R-1340 Wasps on the boats and a single J-5 Whirlwind on the monoplane. Note the Post Office Department delivery vans.

If three are good, four are better or so thought Fokker of America in 1929. The massive F-32 was the 747 of its day, using two pairs of P&W R-1860 Hornet Bs developing 575 hp each. This output had been increased by enlarging the cylinder bore and piston rod. Location of the rear engine caused problems with overheating due to lack of proper airflow over the cylinders. Nonetheless, the F-32 had all the interior appointments one could wish for and resembled a steamship or private rail car in its 1920s Art Deco furnishings. The airplane's career was cut short due to the limitations of its engines but was a noble effort once again by America's pioneering young airline, Western.

An F-32 is at Grand Central Air Terminal in Glendale, California, an airport that figured prominently as the 1930s unfolded. Notice the new National Advisory Committee on Aeronautics (NACA) ring cowls encasing the Hornets for streamlining drag reduction. They did not contribute anything positive to the overheating problems with the rear engines. The sheer size of this airplane cannot be overstated when considering how young the airline business was in 1930. You wonder what is going on with that buckled passenger door! As for engine technology, customers preferred to buy the Pratt R-1830 Twin Wasp instead, which in time became the most numerous aircraft engine ever produced. (Photo Courtesy Gerald Balzer Collection)

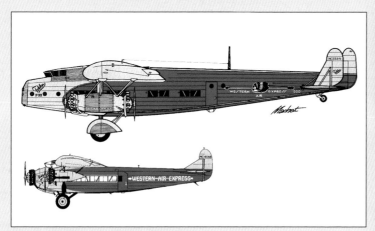

Beautiful scale profile illustration comparing the two Fokkers of their day at Western Air Express: the gigantic F-32 and "normal size" F-10. (Drawings Courtesy Mike Machat, all rights reserved)

Here you can see just how long the cabin of the F-32 was. And how about those velvet seat covers! (Photo Courtesy Gerald Balzer Collection)

Definitely a one-of-a-kind, this 80A-1 at Boeing's Museum of Flight in Seattle provides a rare glimpse at the colors used by Boeing Air Transport (BAT) on its airplanes. Note the Hornet engine's collector ring for cylinder exhaust, which is then channeled back and underneath the wing to move the noise and distracting flames away from the cockpit. Boeing's 80A, "The other trimotor," followed convention of the day combining metal structure at the nose with the steel tube and fabric fuselage. Boeing not only manufactured airplanes but owned an airline and eventually an engine company (Pratt & Whitney), all under the conglomerate umbrella of United Aircraft. BAT was created to operate CAM 18 between San Francisco and Chicago. (Photo Courtesy Wikimedia Commons)

This is a Fairchild 100 Pilgrim in the markings of American Airways. As with most airplanes of the 1930s, the Pilgrim had key metal construction sections mixed with a tube and fabric overall design. Despite its fish-like appearance it was a good airplane, several of which found their way to long service in the Alaska bush. The single engine was a new Wright Cyclone (see chapter 2). The Civil Aeronautics Agency eventually forbade airliners with fewer than two powerplants, which doomed the Pilgrim as well as the exquisite Vultee V-1.

An 80A in flight, showing the three vertical stabilizers that differentiated it from the earlier model of the airplane. (Photo Courtesy AAHS)

The last hurrah for the biplane airliner/mail transport was Boeing's 40C with accommodations for two passengers inside the fuselage just forward of the pilot (who was still stuck in an open cockpit!), and compartments for up to 1,200 pounds of mail. The airplane was powered by a 425-hp Wasp engine. Restoration of this rare aircraft is the handiwork once again of the Pembertons of Spokane, Washington. (Photo Courtesy Addison and Ryan Pemberton with permission)

Hardware on the Wing

With few exceptions, an aircraft is only as good as its engine. A phenomenal powerplant can overcome various deficiencies of a mediocre aircraft, and an unreliable and underperforming engine can render the best airframe useless. Nearly all engines were notoriously unreliable at the start of air travel. As experience was gained and technology progressed, engines became dependable and ushered in a new era of transportation by air.

Throughout the history of aviation, aircraft have relied almost exclusively on engines built specifically for air travel. The Wright brothers made the world's first controlled and powered flight on 17 December 1903. With the help of their mechanic, Charlie Taylor, the Wrights designed and built a unique engine specifically for their aircraft because no engine available had the combination of power and light weight to make the Wright Flyer fly. The Wright engine was a 4-cylinder that had a continuous output of 12 hp and weighed 170 pounds. Although the Wrights used a water-cooled engine, other aviation pioneers used air-cooled engines.

Water versus Air Cooling

A water-cooled engine uses water flowing through jackets or passageways around its cylinders to dissipate heat generated from the engine's combustion process. The water is then cooled in a radiator and returned to the engine. The radiator is exposed to the air slipstream for cooling and creates a lot of drag, especially if the radiator's installation is poorly designed.

In addition to the drag of the radiator, other drawbacks of water-cooled systems include their weight and the often-complex piping, with many points of potential failure, that takes the coolant to and from the radiator. The weight of a water-cooled system includes the radiator, the piping, and the required coolant. The cylinders of water-cooled engines are often arranged in a line or in a V formation. Because of their configuration, water-cooled engines are narrow and have a relatively small frontal area, making them streamlined for installation.

As technology progressed, water-cooling systems were pressurized to prevent the water from boiling. Eventually, ethylene glycol was added to the water to create a mixture that resisted boiling and freezing and proved to be a better thermal conductor. Engines using a water and ethylene glycol mixture were referred to as liquid-cooled engines.

An air-cooled engine uses fins cast or machined into the cylinders to cool the engine. Air flows through the fins as the aircraft flies, and the heat from the engine is transferred into the slipstream. Air-cooled engines can be simpler and lighter than liquid-cooled engines. The cylinders of an air-cooled engine are often arranged around the crankshaft in a radial design. This configuration maximizes each cylinder's exposure to the cooling air, but it usually results in a relatively large engine diameter and much more frontal area than with a liquid-cooled engine. The radial-engine installation often creates more drag when compared to a liquid-cooled engine.

From the early days of aviation through World War I, water cooling seemed slightly more reliable than air cooling, although most engines were generally unreliable. At the time, cylinders were made

Many years of research and experimentation enabled the Wright brothers to make the first powered, controlled, and sustained flight on 17 December 1903. Four flights were made that day in Kill Devil Hills, North Carolina. (Photo Courtesy LOC 00652085)

The engine is mounted atop the lower wing, attached to the wing ribs just below its feet. You can clearly see the chain to flywheel to pulleys to propellers interconnect as well as incoming water-cooling line and gasoline feed line. (Photo Courtesy National Archives via Craig Kodera)

How do you determine when the very first "round" engine came to fruition? These were the *Jurassic Park* days of engine development, both for the new aeroplane as well as the automobile.

The first engine of this design was indeed a radial layout but with water cooling! It was C. M. Manly's 5-cylinder engine of 1901, which produced 52 hp. This engine, when first designed by Stephen Balzer, was actually a rotary engine which, as you might guess, freely spun around the hub thus deriving its cooling. Rotaries were problematic at that time, so water cooling was added to make the whole machine function properly. The Manly-Balzer engine was projected to be used in the Langley Aerodrome, an airplane launched from atop a houseboat in September 1903, trying in vain to beat the Wrights to the First Flight winner's circle. Obviously, Professor Langley had no intentions of making his engine available to any other competing aeronauts, which brings us right back to the Wrights having to design their own.

However, this was not the end of radial-engine development as Jacob Ellehammer created a 3-cylinder design, the world's first true air-cooled airplane engine. He grew the design to five cylinders and used it successfully in 1907 on his triplane. Also in Europe, Alessandro Anzani developed his own 3-cylinder model, which was used to successfully power Louis Blériot's XI monoplane across the English Channel.

The engine engineering world, having not quite cracked the code on air cooling a radial engine, gave us two French design-ers named Georges Canton and Pierre Unne who came up with a successful radial engine in 1909, which was, again, water cooled. This patented engine was offered to and used by the Salmson Company, which produced a useful and important World War I fighter/reconnaissance airplane known as the Salmson 2A2, flown by both France and America during the conflict.

A breakthrough in air-cooling technology occurred (in France once again) with the invention of the first truly successful rotary radial engines known as the Gnome, Clerget, and Le Rhône. The Gnome was the first mass-produced rotary engine, debuting in 1908. Some 80 percent of all World War I aircraft were powered by the Gnome-Le Rhône engines. About 25,000 were manufactured in France while another 75,000 or so were produced by licensees in countries such as Germany, the United States, Austria, Britain, and Sweden. The advantages of air cooling rather than the complexity and weight of liquid cooling allowed for a better weight-to-horsepower ratio for the rotary engine versus the inline.

At the height of its development, the rotary Gnome Lambda produced 200 hp. The more common and reliable Type-N peaked at 160 hp, which is about the high end of everyday rotary engine usage. The superb French Nieuport 28 of 1918 used this engine to great end.

The Manly-Balzer radial engine, which consisted of five cylinders and developed an amazing 52 hp in 1903, is on display at the Smithsonian's National Air and Space Museum. Not only did the world have bicycle-builder aviators in Ohio named Wright, but the cycle-themed wheel makes its appearance here as well; very telling of motive technology at that time. (Photo Courtesy Smithsonian National Air and Space Museum TMS A19080003000)

On the Potomac, Professor Langley's Aerodrome perched upon the houseboat, which housed the intricate and massive launching deck and rail apparatus designed expressly for the flying machine. Two flight attempts proved to be unsuccessful, but the world received radial-engine technology, a more-than-fair trade in most estimations. (Photo Courtesy Smithsonian National Air and Space Museum NASM A-18801)

The 90-hp Curtiss OX-5 was the first aircraft engine mass produced in the United States. First run in 1915, the OX-5 is a typical example of a water-cooled aircraft engine with individual cylinders common for that era. (Photo Courtesy NARA 17339427)

The rotary engine provided a combination of lightness, power, and reliability in the World War I era. The propeller was mounted to the crankcase and the entire engine rotated; the stationary crankshaft was mounted to the aircraft. This French Gnome Monosoupape 9 Type B-2 engine produced 100 hp and was tested by the US Navy. (Photo Courtesy NARA 518849)

The Liberty V-12 was selected as the "universal" aircraft engine for US production during World War I. More than 20,400 examples of the 400-hp 1,450-ci engine were built. However, the Liberty was made too late to see much action during the war. The huge surplus of Liberty engines that existed after World War I slowed development of other types. (Photo Courtesy NARA 518852)

from steel or cast iron, and the cooling fins of air-cooled engines were often thick and few in number. These factors limited the cooling efficiency of the cylinder.

Shortly before World War I, a new type of air-cooled engine gained popularity. These engines were known as rotaries; they had the same radial configuration as other air-cooled engines, but the entire engine rotated. The forward speed of the aircraft combined with the spinning of the engine provided the cylinders with ample air for cooling. Rotary engines had to be limited in size and power because the gyroscopic effects of the spinning engine were already a cause for concern. As engines were designed to be larger and produce more power, rotary engines could not keep up with the competition. The rotary aircraft engine did not progress much beyond World War I.

The first aircraft engine mass produced in the United States was the Curtiss OX-5. The water-cooled V-8 engine had a 4.0-inch bore, a 5.0-inch stroke, and a displacement of 503 cubic inches (ci). The OX-5 produced 90 hp and was first run in 1915. Mass production of the engine started in 1917, in response to the need for trainer aircraft during World War I. Although more powerful and lighter engines existed, the OX-5 was proven and available. Ultimately, more than 12,000 OX-5 engines were produced.

At the start of World War I, the United States lacked a good, front-line aircraft engine and moved to obtain licenses to manufacture some of the more powerful European models. But, converting the engines and their drawings from metric to Imperial units proved to be troublesome, as did attempts at mass-producing engines that were previously built in small numbers. The manufacturing difficulties that accompanied the European designs, combined with the need for a more powerful engine, led the United States to focus on designing and producing an entirely new aircraft engine. This new engine made full use of interchangeable parts and mass production.

Jesse G. Vincent of the Packard Motor Car Company and Elbert J. Hall of the Hall-Scott Motor Car Company met in Washington, DC, on 29 May 1917. Over the next five days, and with the assistance of a few draftsmen, the men designed a series of engines that used interchangeable parts, all of which could be easily mass produced. A short time later, the engine series became known as the Liberty.

All of the Liberty engines used the same 5.0-inch bore and 7.0-inch-stroke cylinders and as many other common components as possible. The Liberty was water-cooled and made of an aluminum crankcase and steel cylinders. Four engines were designed: a 100-hp inline 4-cylinder of 550-ci displacement, a 210-hp inline 6-cylinder of 825-ci displacement, a 270-hp V-8 of 1,100-ci displacement, and a 400-hp V-12 of 1,649-ci displacement. However, only the V-12 entered mass production, with 20,478 examples built.

When the war ended in 1919, so many Liberty V-12 and Curtiss OX-5 engines hit the surplus market that it slowed the development of new engines from established manufacturers. Although the Liberty was fairly expensive, surplus OX-5s were relatively cheap. OX-5 engines continued to be used in newly designed aircraft until the late 1920s. In 1929, 38 percent of all licensed aircraft in the United States were powered by the OX-5, an engine considered outdated when it entered mass production more than 10 years earlier.

An application of the inline engine on a commercial airplane was presented by the designers at Curtiss Aircraft in the late 1920s. The Condor 18 was a civilian version of the B-2 Condor bomber for the Air Corps. Two Curtiss V-1570-7 "Conqueror" liquid-cooled V-12 engines produced the hallmark inline high horsepower of 600 each, far outpacing any radial engine of the time. Water cooling was a high-maintenance aspect of these closed-system engines, with any drag reduction realized by their slim faces lost immediately by those gigantic radiators!

Although surplus World War I engines may have stagnated the development of new types, the war advanced the aviation industry in many ways. Prewar aircraft manufacturing was a cottage industry producing a few aircraft each year. The war transformed the industry into an organized operation using assembly-line processes to mass produce aircraft. Improved machining made the manufacture of complicated parts with tight tolerances possible. The use of aluminum became widespread, especially for engines.

A few years after World War I, the simplicity and reliability of air-cooled radial engines earned them preference over the readily available water-cooled Liberty. In addition, the larger air-cooled engines of the late 1920s offered power well in excess of the 400 hp available from the Liberty. Air-cooled engines were now dependable, and the reliably of the engine and the aircraft were critical components needed to bring air travel to the masses. With the necessary technology and equipment in place, the airliner was born.

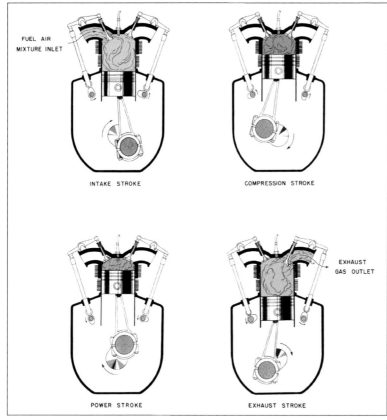

The four-stroke process of the Otto cycle brought about a revolution for the internal combustion engine. A continuing trend of lighter and more powerful engines enabled the widespread adoption of the automobile and made sustained flight possible. (Photo Courtesy **Instruction Manual Pratt & Whitney Engines R-1830-43 & 65, November 1943)**

Engine Power

Nearly all piston aircraft engines operate on the Otto cycle, four-stroke principle, first made practical by Nikolaus Otto in 1876. Each up or down movement of the piston is a stroke, and the crankshaft makes two revolutions to complete a cycle.

The first stroke is the piston moving down in the cylinder and drawing in air and fuel through an open intake valve. The second stroke is the piston moving up to compress the air and fuel charge in the cylinder with all valves closed. The crankshaft has made one revolution at the end of the second stroke. When the piston is near its highest point in the cylinder, top dead center (TDC), a spark plug fires and ignites the volatile air and fuel mixture. The rapid expansion of the combusting gases in the cylinder force the piston back down on its third stroke. On the fourth stroke, the exhaust valve has opened and the piston moves up in the cylinder to push out the exhaust gases in preparation for the cylinder to restart the cycle. The second revolution of the crankshaft has now been completed. The four strokes are known as intake, compression, power, and exhaust.

In the Otto cycle, each cylinder produces only one power stroke for every two revolutions of the crankshaft. Half of the engine's cylinders fire during one revolution of the crankshaft, and the other half fire during the next revolution. The firing order is spaced to smooth out the power pulses created by the power strokes. An 8-cylinder engine fires a cylinder every 90 degrees of crankshaft rotation; a 9-cylinder engine fires every 80 degrees; and an 18-cylinder engine fires every 40 degrees.

Engine Configuration

Engines have been made in many different configurations, but the most common during the piston-engine's golden age of flight were the V and radial. Engines configured in a V need to have an even number of cylinders. Four-stroke radial engines require an odd number of cylinders on each row: an 18-cylinder radial has two rows of nine cylinders. These configurations keep the firing order of the cylinders even, so that every other cylinder fires. That firing order can only be achieved with an odd number of cylinders per row.

Many factors contribute to how an engine makes power. The most basic are the bore and stroke of the cylinder and the engine's RPM. In general, the larger the bore and the longer the stroke, the more power the cylinder produces. However, there are practical limits to how large a cylinder can be built, as stresses on engine components increase with their size. The faster the engine speed (RPM), the more power the engine produces. But there are limitations on RPM also; engine components can only move so fast, and the stress on those components increases with speed. Adding more cylinders can make an engine more powerful, but again, there are practical limits. As the engine grows in size and weight, its installation in an aircraft becomes increasingly difficult.

Options are available for increasing an engine's power while keeping its basic structure unchanged. Chief among these options are increasing the engine's compression ratio and supercharging the engine with higher boost pressures. However, these options require using fuels compatible with the higher pressures inside of the cylinder. Compression ratio, boost pressure, and fuel quality are variables that directly affect one another, and as technology progressed, it became easier to employ these elements to produce more power from the engine.

Compression ratio is the relative difference in cylinder volume between when the piston is at its lowest point (bottom dead center, BDC) in the cylinder and when the piston is at its highest point (TDC). If the volume above the piston at TDC is 20 percent of the volume when the piston is at BDC, the cylinder has a 5:1 compression ratio. If the volume above the piston at TDC is 10 percent of the volume when the piston is at BDC, the cylinder has a 10:1 compression ratio. An engine with a 10:1 compression ratio produces around 32 percent more power than an engine with a 5:1 compression ratio.

An engine and its components must be designed to resist the greater cylinder pressures that accompany a higher compression ratio. In addition, higher compression raises the temperature inside of the cylinder. If the engine is not properly designed for higher compression ratios, a hot spot can occur inside of the cylinder. The hot spot can cause the uncontrolled ignition of the air and fuel mixture before the compression stroke is completed and the spark plug has fired. This is called pre-ignition, and it can quickly damage the piston, valves, and other engine components. Pre-ignition can increase the pressure inside a cylinder enough to bend the connecting rod that attaches the piston to the crankshaft. The quick and substantial damage caused by pre-ignition can destroy the entire engine in seconds.

As engines advanced and were developed to produce more power, the exhaust valve was a common source for pre-igniting the incoming air and fuel charge. Methods of cooling the exhaust valve were investigated, and the most successful solution was to use a sodium-cooled valve (developed in the mid-1920s by Sam Heron). The stem, and often the head, of a sodium-cooled valve are hollow and partially (approximately 2/3) filled with sodium. Once the valve reaches 208 degrees Fahrenheit, the sodium melts. The up-and-down movement of the valve sloshes the sodium up and down inside the valve. The sodium picks up heat from the valve head, cooling it, and transfers that heat to the valve stem. The valve stem extends outside

The complexity of radial engines is illustrated here by all the parts laid out for inspection. On each table are the components for a 14-cylinder P&W R-1830 Twin Wasp. (Photo Courtesy LOC 2017694151)

of the cylinder and transfers the heat to the valve guide boss and subsequently to the cylinder cooling fins.

One of the first applications of sodium-cooled valves was on the Wright J-5 engine installed in the *Spirit of St. Louis*. After its successful transatlantic flight, use of sodium-cooled valves spread to air racers and then to standard engines. In later engines, the valves contained a sodium and mercury mixture that was liquid at room temperature.

Fuel

As the cylinder's compression ratio is increased, the performance characteristics of the fuel must be able to handle the higher compression. When the spark plug fires, the fuel in the cylinder needs to ignite and burn progressively throughout the cylinder, away from the spark plug. The fuel burns; it does not explode all at once. The flame front travels around 35 feet per second when combustion begins. That speed increases to about 150 feet per second before slowing as the volatile air and fuel mixture inside the cylinder is consumed.

If underperforming fuel is used in conjunction with a high compression ratio, pockets of fuel in the cylinder may self-ignite before the flame front of the burning fuel reaches the pockets. This phenomenon is called detonation, and it is essentially a series of uncontrolled explosions occurring in the cylinder after the spark plug has fired. When the supersonic shock waves created by detonation collide with the cylinder walls, a pinging or knocking sound results. Because of this sound, detonation is often called pinging or knocking. Detonation damages and erodes the piston and cylinder head, and it can cause heat significant enough to melt part of the piston or even burn through the piston's head. Once a hole has been created in the piston, high-pressure combustion gases can pressurize the crankcase and combustion products can enter the oil system. In many cases, the onset of detonation is not noticed until parts have been significantly damaged.

Fuel with a higher octane rating or performance number can handle higher compression ratios and resist detonation. During World War I, American aviation fuel had an estimated octane rating of around 50 (octane ratings did not exist until the late 1920s). In the 1920s, it was known that fuel from different locations had different properties. High-powered engines ran much better on fuel from California (75 octane) than on fuel from Pennsylvania (55 octane). In addition, it was discovered that adding small amounts of tetraethyl lead (TEL) enhanced the fuel's performance. The differences in fuels and their enhancement with additives led Graham Edgar to investigate fuels and develop the octane rating system in 1926.

The octane rating was devised by analyzing various mixtures of heptane and isooctane to determine the mixture's ability to resist knock (detonation). In its pure form, heptane has no resistance to detonation and burns explosively. Because of these characteristics, heptane was given an octane rating of 0. In its pure form, isooctane has a good resistance to detonation and burns smoothly. Because of these characteristics, isooctane was given an octane rating of 100. A 50/50 mixture of heptane and isooctane has an octane rating of 50. The scale is applied to other fuel mixtures, and an octane rating is assigned based on the fuel's anti-detonation performance.

When the fuel's ability to resist detonation exceeds 100 octane, a performance number (PN) is used. The PN corresponds to the fuel's ability to resist detonation compared to 100 octane. For example, 120-PN fuel resists detonation 20 percent more than 100-octane fuel. There is no octane greater than 100. However, many people still say "octane" when referring to fuels with a PN higher than 100.

By the 1930s, aviation fuel was around 87 octane, and fuel was one of the factors limiting higher power aircraft engines. In 1934, Shell made the first batch of 100-octane fuel, and the fuel was mass produced by 1938. Fuel development ultimately led to the 100/130 PN and 115/140 PN fuels created during World War II. The split numbers represent the fuel mixture's knock rating under lean/rich conditions.

Fuel's resistance to detonation allows for a higher compression ratio and/or a higher degree of supercharging of the incoming fuel and air mixture. As the piston moves down on the intake stroke, a vacuum is created that draws in air through the intake valve. If the air is being drawn in at atmospheric pressure, the engine is normally aspirated. Devices can be fitted to an engine to pressurize the incoming air charge and force it into the cylinder. The more air and fuel that can be crammed into the cylinder, the more power that cylinder makes. Providing the incoming charge at pressure levels higher than atmospheric is known as supercharging.

Supercharging

In the 1920s, radial engines began to employ blowers as a way to mix the air and fuel and distribute the mixture evenly to all the cylinders. The blower was essentially a supercharger that did not provide much boost. As engine development progressed, the blower was made to provide more and more boost, becoming a true supercharger. With the exception of a few experimental models, all Pratt & Whitney (P&W) engines had an integral, mechanically driven blower/supercharger mounted in the rear engine section. Wright began to incorporate integral, mechanically driven blowers/super-

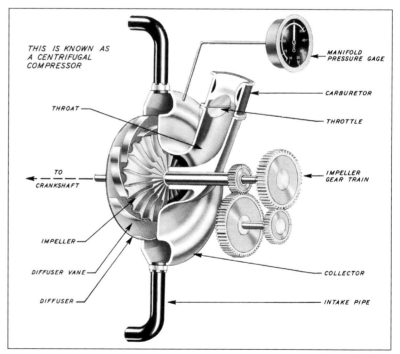

THIS IS KNOWN AS A CENTRIFUGAL COMPRESSOR

MANIFOLD PRESSURE GAGE

CARBURETOR

THROTTLE

THROAT

TO CRANKSHAFT

IMPELLER GEAR TRAIN

IMPELLER

DIFFUSER VANE

COLLECTOR

DIFFUSER

INTAKE PIPE

*Nearly all superchargers for aircraft engines, especially radials, employed a centrifugal compressor (impeller). The impeller was driven by the crankshaft via step-up gears, often at six or eight times crankshaft speed. The impeller's tips were kept below supersonic speeds. (Photo Courtesy **The Aircraft Engine and Its Operation**, December 1952)*

chargers on its production engines in the late 1920s, starting with the J-6 Whirlwind and R-1820 E Cyclone.

Two types of devices can be used to "supercharge" air into the cylinder: a mechanically driven supercharger and a turbosupercharger (or turbocharger, or turbo). With both superchargers and turbochargers, the pressurized air created by the device is fed into the engine's intake manifold. Engines can have multistage and multispeed superchargers or a combination of a supercharger and a turbocharger. Because the incoming charge is under pressure, highly supercharged (boosted) engines generally run a lower compression ratio than normally aspirated (non-boosted) engines. In fact, some engines had their compression ratio decreased as a new supercharger with a higher degree of boost was incorporated.

An engine can be so highly supercharged that it destroys itself if operated at full power at low altitudes. Such power may seem counterproductive, but it is essential in the thinner air at higher altitudes. Although the loss is not linear, a normally aspirated engine loses approximately 3 percent of its power with every 1,000 feet of altitude. If the engine is rated for 1,000 hp at sea level, it produces around 738 hp (26 percent loss) at 10,000 feet and 448 hp (55 percent loss) at 25,000 feet. A high degree of supercharging allows the engine to continue to produce its rated power at these higher altitudes.

At a certain altitude, even a supercharged engine begins to lose power. This altitude is called the critical altitude, and it is a function of the supercharger's level of boost. An engine with a higher degree of supercharging can have a higher critical altitude.

A mechanically driven supercharger is a pressure blower typically mounted to the back of the engine, but it can be mounted remotely as well. The supercharger's impeller (or compressor, or rotor) is driven by the engine via a series of gears and clutches or a hydraulic coupling. Driving the supercharger consumes a certain amount of engine power, and supercharger operation on larger engines can require a few hundred horsepower. From 1930 on, nearly all radial engines had a mechanically driven supercharger.

A two-speed supercharger enables the engine and aircraft to have two critical altitudes. Essentially, the supercharger's impeller is shifted from one speed to another as altitude is gained. For example, the supercharger's low-speed gear enables the engine to generate peak power up to 15,000 feet; the high-speed gear enables the engine to generate peak power up to 25,000 feet.

Two-stage supercharging uses two compressors, with the first stage feeding pressurized air to the second stage. Mechanically driven, two-stage superchargers on radial aircraft engines employ two impellers. The first stage is typically an auxiliary stage added to the engine and only engaged above a certain altitude. Once engaged, the auxiliary stage feeds pressurized air to the main impeller that is integral with the engine. Although it is technically the second stage of supercharging, the integral impeller is often referred to as the main stage, because it is always in operation.

If the first stage of supercharging has a pressure ratio of 2.5:1 and the second stage has a pressure ratio of 2.0:1, the complete two-stage supercharger has a pressure ratio of 5.0:1. A two-stage supercharger can pressurize the air much more efficiently than a single-stage unit trying to match the same 5.0:1 pressure ratio. Keeping with the 5.0:1 pressure ratio example, the supercharger is not used to boost sea-level air pressure of 14.7 psi (30 inches of mercury) to 73.5 psi (150 inches of mercury), but it is used to boost the 5.5 psi (11 inches of mercury) of air pressure at 25,000 feet to 27.3 psi (56 inches of mercury).

A turbosupercharger is a remote blower driven by the engine's exhaust gasses. Exhaust traveling at high speed turns a turbine wheel inside the turbocharger. A compressor wheel (or impeller wheel) is mounted on the same shaft as the turbine wheel but in a different section of the turbocharger. The compressor wheel pressurizes the air that is fed into the engine's intake. Turbochargers are lighter than mechanically driven superchargers, and the turbocharger's operation consumes very little engine power. In addition, turbochargers can provide more boost at higher altitudes than superchargers. However, the construction of turbochargers requires exotic materials with a resistance to high temperatures, and the turbocharger can have complex controls. In radial engines, turbochargers are typically added as the first (auxiliary) stage of supercharging, providing pressurized air to the engine's mechanically driven supercharger, which serves as a second (main) stage.

The high pressure ratio of a two-stage mechanically driven supercharger, or a turbocharger/supercharger combination, increases the temperature of the incoming charge to unacceptably high levels. To cool the incoming charge, either an intercooler or aftercooler is used. An intercooler is positioned between the first and second supercharger stage; an aftercooler is placed between the second stage and the engine. The intercooler or aftercooler acts as a heat exchanger, transferring heat from the incoming charge to either air or water flowing through the cooler.

Although used extensively in many different types of aircraft during World War II, two-stage supercharging and turbochargers were rarely used in commercial aviation. The only application of turbochargers in commercial aviation was on the P&W R-4360 engines in the Boeing 377 Stratocruiser. Mechanically driven, single-stage superchargers were more than adequate for the airliners' operating altitudes, which were typically around 25,000 feet for pressurized aircraft.

An example of how better fuel and greater supercharging can enhance engine performance is found in an experiment conducted

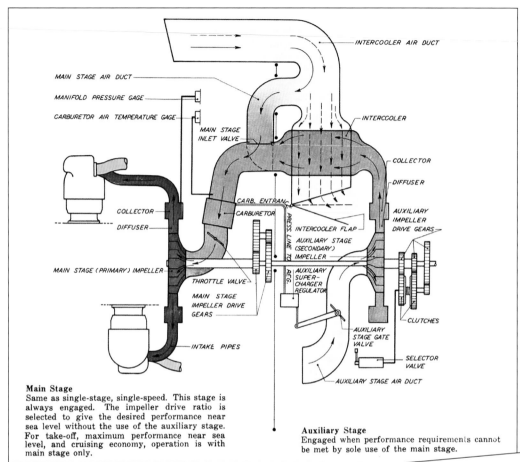

One form of two-stage supercharging employed two impellers. The air pressurized from the first impeller passed to the second impeller, where it was further pressurized. An intercooler was typically installed between the stages to cool the air. Some engines used an aftercooler, which was positioned between the final stage of supercharging and the engine. A clutch and gear set allowed the supercharger to be driven at two or even three speeds. In later engines, the impeller (typically of the axillary stage) was driven by a fluid coupling that enabled variable speeds. (Photo Courtesy *The Aircraft Engine and Its Operation, December 1952*)

Main Stage
Same as single-stage, single-speed. This stage is always engaged. The impeller drive ratio is selected to give the desired performance near sea level without the use of the auxiliary stage. For take-off, maximum performance near sea level, and cruising economy, operation is with main stage only.

Auxiliary Stage
Engaged when performance requirements cannot be met by sole use of the main stage.

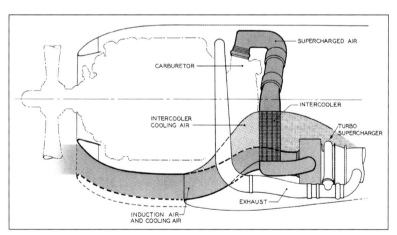

A typical turbocharger installation involved ducting the exhaust gases from the engine to the remotely mounted turbocharger. Once the incoming air had been pressurized by the turbocharger, it passed through an intercooler to decrease its temperature. The air was then fed into the supercharger integral with the engine, where the air was further pressurized before being ducted to the cylinders. (Photo Courtesy *Installation Handbook, March 1948*)

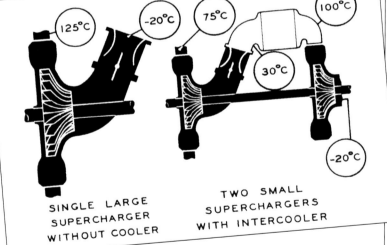

Compared to a two-stage supercharger, a single-stage supercharger was less efficient and generated a larger increase in induction air temperature. Of course, an aftercooler could have been added to the single-stage supercharger, but the two-stage unit was overall more efficient and more versatile. (Photo Courtesy *The Aircraft Engine and Its Operation, December 1952*)

by Sam Heron in 1931. A 450-hp P&W R-1340 Wasp engine was run on a special blend of California fuel that had been further enhanced with a small amount of TEL. The fuel concoction had an octane rating of 98. Modified supercharger gears were installed that made the supercharger impeller spin at a higher RPM, boosting the air fed into the engine. The special fuel and higher manifold pressure caused the engine's output to double to 900 hp. The extra power created reliability issues with other engine components, but the test gave insight into the relationship between high-quality fuels and high boost pressures and the high-power outputs that could result.

A higher compression ratio allows an engine to make more power, but the compression ratio must be matched to the strength of the engine and the quality of the fuel. Higher octane/PN fuel can allow an engine to make more power, but the fuel must be matched to the compression ratio and pressure of the incoming air charge. Supercharging the incoming air charge can make more power, but fuel, compression ratio, and the engine's (or aircraft's) critical altitude must be considered. All of these variables affect one another and must be taken into account by the engine's designer. During the 1920s, all of these variables and their interconnected relationship were just beginning to be understood, making the engine-design process that much more difficult.

Around 1930, engines began to have two power ratings, one for takeoff and one for normal operation. The takeoff rating was intended for a short duration of 2 to 5 minutes. Running at increased power for longer durations risked damage to the engine. The takeoff was the most crucial moment of flight and required the most power. Allowing engines to "overproduce" for a short period of time made the takeoff safer, as it got the aircraft off the ground and to a safe altitude. The normal rating was the power the engine could produce for an unlimited amount of time. No engine damage would occur from the engine operating for hours at its normal rating.

Later, two other ratings were established, maximum continuous power and military or emergency power. Maximum continuous power was typically the same as normal power, but in some cases, it was more power than the engine's normal rating. It could cause extra wear and decrease the time between engine overhauls, but maximum continuous power provided additional output and safe operation when needed. Another term used was maximum except

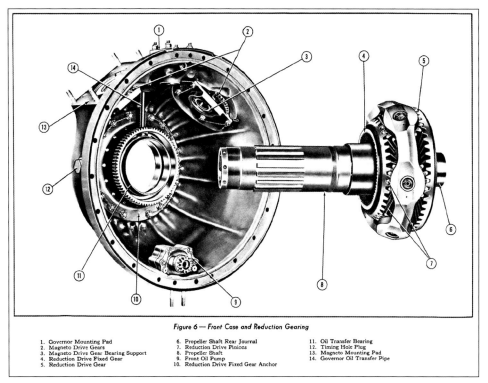

Figure 6 — Front Case and Reduction Gearing

1. Governor Mounting Pad
2. Magneto Drive Gears
3. Magneto Drive Gear Bearing Support
4. Reduction Drive Fixed Gear
5. Reduction Drive Gear
6. Propeller Shaft Rear Journal
7. Reduction Drive Pinions
8. Propeller Shaft
9. Front Oil Pump
10. Reduction Drive Fixed Gear Anchor
11. Oil Transfer Bearing
12. Timing Hole Plug
13. Magneto Mounting Pad
14. Governor Oil Transfer Pipe

The bevel gear reduction on the P&W R-1830 had the crankshaft coupled to and turning a driving gear (6), which turned six beveled pinions (7) held in a carrier integral with the propeller shaft (8). A fixed gear (4) forced the beveled pinions and propeller shaft to turn at a reduced speed compared to the crankshaft. (Photo Courtesy Service Instructions R-1830-75 and -98, February 1945)

takeoff (METO) power. Military or emergency power was very similar to the takeoff rating.

Propellers and Gear Reduction

Although an aircraft may only be as good as its engine, an engine is only as good as its propeller. All the power produced by the engine must be converted into thrust to propel the airplane. In the early days of aviation, nearly all propellers were made of wood. As engine power increased, wood propellers were prone to flexing. Other issues arose when wood propellers became damaged or even delaminated while flying through rain. Starting in the 1920s, a switch to metal (aluminum and steel) propellers solved these issues, but more trouble lay ahead.

Engines were often direct drive, meaning the propeller was attached to the crankshaft and turned at the same RPM as the engine. But a propeller can only spin so fast. As the tips of the blades approach the speed of sound, the propeller's efficiency drops off drastically. As engines became more powerful and aircraft became bigger, a larger propeller was needed. The larger the propeller, the slower it needs to be turned. An example of the problem is when the engine creates its optimum power at 3,000 rpm, but the propeller's optimum speed is 2,000 rpm. The solution is to add a gear reduction between the engine and the propeller. This allows the engine to turn at 3,000 rpm but turns the propeller via a .667 reduction at 2,000 rpm. Having a propeller reduction gear enabled both the engine and the propeller to run at their optimal speeds.

Toward the end of the piston-engine era, the gear reduction ratio was selected after the engine and aircraft were designed. An analysis occurred to find the optimum propeller size and speed for the engine and aircraft combination with the intended purpose of the aircraft in mind. With the propeller specifics known, a gear reduction was then designed for that propeller. In addition, some engines, such as the Wright R-3350, could be fitted with different nose cases that housed different propeller gear reduction speeds. This allowed the same basic engine to be used in different aircraft types by simply selecting different nose cases.

Metal propeller blades and engines with gear reduction solved many issues, but another major limitation of early flight was the pitch of the propeller blade. Early propellers were fixed-pitch, mean-

ing that the pitch could be adjusted only when the aircraft was on the ground. The pitch of the propeller blade affects its performance at various stages of flight. A relatively flat pitch (or angle) is good for acceleration; a coarser pitch is good for an efficient cruise speed. Good acceleration is needed for takeoff, but a long range and a high cruise speed are paramount for air travel. A fixed-pitch propeller was a compromise between these equally desirable characteristics. The coarsest pitch that still allowed the aircraft to safely takeoff was often chosen. This resulted in poor takeoff performance and poor cruise performance.

An example of the fixed-pitch propeller issue is illustrated by Charles Lindbergh's historic flight across the Atlantic. Because the aircraft's cruise economy was so important, the ground-adjustable propeller was set half a degree off from its ideal cruise angle. This resulted in poor acceleration, and the *Spirit of St. Louis* barely got airborne from Roosevelt Field on 20 May 1927. The aircraft cleared telephone lines by 20 feet, and Lindbergh may have been perfectly happy to fly under them. However, such a narrow margin of error is not acceptable for passenger airline travel.

The solution to the limitations of a fixed-pitch propeller was to create an adjustable-pitch propeller, with which the pilot could change the propeller's blade angle during flight. A flat pitch could be used for the takeoff and climb, and then the pitch could be made

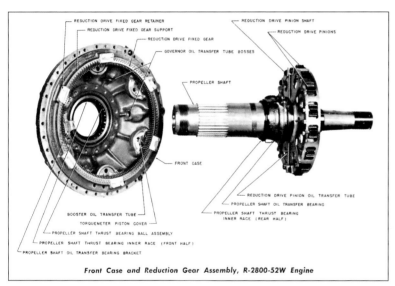

The planetary (epicyclic) gear reduction used on the P&W R-2800 C-series had a gear (not pictured) coupled to the crankshaft. The gear engaged the inner sides of the reduction drive pinions. The pinions were held in a carrier integral with the propeller shaft. The outer side of the pinions engaged the reduction drive gear fixed to the front nose case. The interaction between the fixed gear and the pinons caused the propeller shaft to turn at a reduced speed compared to the crankshaft. (Photo Courtesy Handbook Overhaul Instructions Models R-2800-50, -52W, -54, -99W, -103W Aircraft Engines, April 1956)

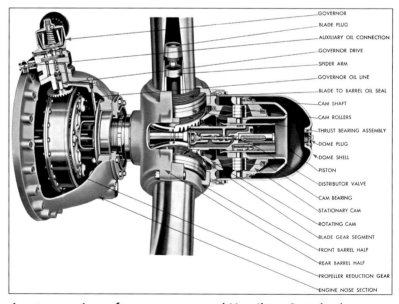

A cutaway view of a constant-speed Hamilton-Standard propeller illustrates the complicated mechanisms that varied the blade angle depending on oil pressure. Any propeller that was adjustable in flight increased the efficiency of converting engine power to thrust. (Photo Courtesy Aircraft Engine Historical Society)

A Sikorsky S-40 dubbed American Clipper *by PAA flies over the Hudson River, separating New York and New Jersey, in October 1931. Townend rings surround the aircraft's P&W R-1690 Hornet engines. The Townend ring was an improvement over exposed cylinders but was not as efficient as the NACA cowling. (Photo Courtesy William Pearce Collection)*

The Curtiss AT-5A aircraft was modified to test the efficiency of the NACA cowling. On the left is a standard AT-5 with the cylinders of its Wright J-5 engine exposed. It is easy to see how this configuration created a lot of drag. On the right, the NACA cowling has been constructed to aerodynamically enclose the engine and direct airflow through its cylinders. The fuselage was modified to create a smooth transition between it and the cowling. The modifications increased the aircraft's speed by 16 percent. (Photo Courtesy NASM CW8G-T-3375 [left] and NASA EL-2000-00345 [right])

coarser as the aircraft picked up speed and transitioned to cruise flight. A flat pitch could be used to help the aircraft decelerate as it slowed for its landing approach. The blades of some propellers could also be reversed to help brake the aircraft after touchdown. The adjustable-pitch propeller was developed in the late 1920s and early 1930s, and for its era, it was just as revolutionary to air travel as the jet engine.

Engine Cowlings

As mentioned previously, the drag of an air-cooled engine can be significant. A water-cooled engine can be enclosed in a streamlined cowling, but an air-cooled engine needs airflow through its cylinders for cooling. Originally, air-cooled engines were installed in aircraft so that their cylinders were exposed to the slipstream, which created a tremendous amount of drag (resistance). Drag increases at the square of speed. The quadratic increase in drag means that the power required to overcome the drag and propel the craft faster increases at the cube of speed. All things being equal, going 20 percent faster creates 44 percent more drag and requires 73 percent more power.

Reducing drag has the double benefit of increasing an aircraft's speed and extending its range without any increase in engine power. If drag reduction increases an aircraft's speed from 100 to 120 mph, and the aircraft has five hours of fuel, the range has also increased from 500 to 600 miles.

Reducing the drag of air-cooled engines became a major focus in the late 1920s. Hubert C. Townend of the National Physical Laboratory in England began studying ways to decrease drag in 1927. He developed a ring that went around a radial engine, surrounding its cylinders. The ring had an airfoil cross-section and was known as the Townend ring. It decreased drag by smoothing the airflow around the engine and its cylinders.

Townend's efforts were unknown to Fred Weick, who was working on the same issue for the National Advisory Committee on Aeronautics (NACA). Weick also began his research in 1927 but expanded it to include how the engine's installation interacted with the airframe. Weick developed the NACA cowling, which completely enclosed the engine and provided a streamlined exit for the cooling air. Tests conducted in October 1928 on a Curtiss AT-5A trainer demonstrated an increase in top speed from 118 mph with the exposed engine to 137 mph with the cowling. In 1929, the Lockheed Vega transport's top speed increased from 165 mph with an exposed engine to 190 mph with the NACA cowling. In 1932, the prototype for the Martin B-10 bomber was switched from Townend rings to NACA cowlings and achieved a top speed increase from 195 mph to 225 mph.

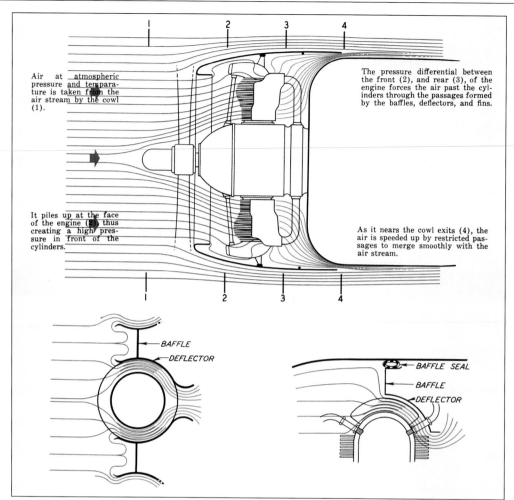

Air at atmospheric pressure and temparature is taken from the air stream by the cowl (1).

The pressure differential between the front (2), and rear (3), of the engine forces the air past the cylinders through the passages formed by the baffles, deflectors, and fins.

It piles up at the face of the engine (A) thus creating a high pressure in front of the cylinders.

As it nears the cowl exits (4), the air is speeded up by restricted passages to merge smoothly with the air stream.

BAFFLE
DEFLECTOR

BAFFLE SEAL
BAFFLE
DEFLECTOR

Engine cowlings began as a way to streamline the engine's installation and increase the aircraft's speed. The cowling, combined with numerous baffles, developed into a sophisticated way to cool the engine and minimize drag by routing the air where it was most needed. (Photo Courtesy The Aircraft Engine and Its Operation, *December 1952)*

As NACA cowling designs progressed, the fixed exit slot for the cooling air was replaced by exit doors called cowl flaps. Cowl flaps could be opened and closed to moderate the amount of cooling air that flowed through the cowling and were used to maintain a fairly constant engine temperature. The cowl flaps could be open for ground operations, slightly closed for takeoff and climb, and mostly closed for cruise. Cowl flaps were kept as closed as possible to offer the least amount of drag.

With the engine encased in a fully enclosed cowling and cowl flaps allowing the minimum amount of air required to cool the engine, ever more complex baffles were used to direct the flow of cooling air where it was needed inside of the cowling. Initially, the air that entered the cowling was sufficient to cool the cylinders as it found its own way out. As engines developed more power, the air needed to be directed to flow through the evermore finely pitched cooling fins. To reduce drag to a minimum and to provide the most efficient cooling, all of the air that entered the cowling needed to flow through the cooling fins. The fit of the baffles became crucial

to keeping high-powered radials cool under prolonged operation. The baffles often incorporated blast tubes that directed a stream of cooling air at the spark plugs to prevent their ignition leads from overheating. Designing streamlined cowlings and baffling that efficiently cooled the engine and reduced drag became an engineering art form.

Wright Aeronautical History up to the Whirlwind (1930)

The history of the Wright Aeronautical Corporation began with the Wright brothers. Following their 1903 flight, Orville and Wilbur Wright continued to refine their flying machine. They received very little attention until around 1905, but by 1908, they had proven themselves to be masters of the air.

Backed by bankers and businessmen, the Wright Company was formed in New York City, New York, on 22 November 1909 to manufacture aircraft. The Wright brothers continued to design and develop their aircraft and aircraft engines, but a number of com-

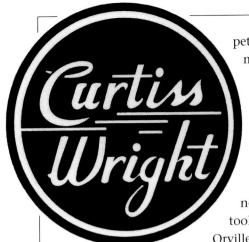

petitors had entered the aviation marketplace. The Wrights filed numerous lawsuits to protect their aircraft patent against infringement. To emphasize the patent's legitimacy, the Wrights were hesitant to adopt new features on their aircraft.

Wilbur Wright died of an illness on 30 May 1912, and Orville took over as the company's president. Orville did not have the same business acumen as his brother, and he knew it. By 1915, aircraft development at the Wright Company had fallen far behind that of its competitors. Orville decided it was time for a change. He bought out the original investors and sold the Wright Company to investment bankers in October 1915.

The company's new management felt that the conflict in Europe would provide a sales opportunity and moved quickly to produce a product. A license to build the French Hispano-Suiza 8Aa engine was obtained in January 1916. The 8Aa was a water-cooled V-8 aircraft engine with an output of 150 hp at 1,450 rpm. As produced by Wright, the engine was known as the Model A, and it was manufactured at a newly equipped plant in New Brunswick, New Jersey. France immediately placed an order for 450 engines.

In August 1916, the Wright Company merged with the Glenn L. Martin Company to form the Wright-Martin Aircraft Company. This union was short-lived, as management from the Wright Company wanted to focus on aircraft engine production and was not eager to manufacture aircraft for which the Martin plant in California did not have the capacity to produce. Glenn Martin resigned in 1917 and started another company under his own name.

By July 1917, fewer than 100 Model A engines had been delivered to France. However, engine production rates were increasing rapidly, and the United States placed an order for 500 engines. An improved version of the Model A, the Model I, entered production, along with the larger and more powerful Model E and Model H. The Model E produced 180 hp at 1,800 rpm, and the Model H produced 300 hp at the same RPM. By the end of the war in 1918, Wright-Martin had delivered 5,816 engines. Production was still expanding, and the company had planned to produce 2,000 engines a month by mid-1919. These plans were never realized.

At the end of World War I, there were three major aircraft engine manufacturers in the United States: Curtiss, Packard (building Liberty engines), and Wright-Martin. With the exception of a few experimental models, all of the engines produced by these compa-

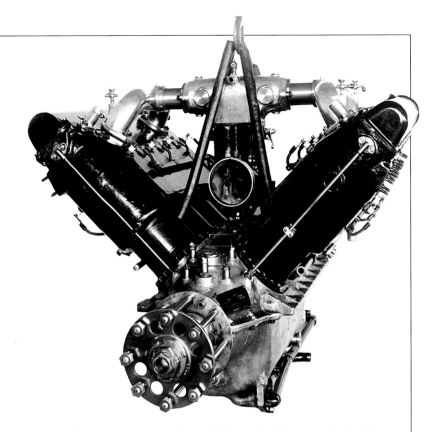

The Wright H engine was derived from the Hispano-Suiza 8A. Crude by today's standards, the engine was at the forefront of technological development for its time. With is monobloc cylinder construction, the engine foreshadowed future liquid-cooled engines such as the Curtiss D-12 and the Rolls-Royce Merlin. (Photo Courtesy Aircraft Engine Historical Society)

nies were water-cooled Vs.

The management at Wright-Martin continued to focus on aircraft engines in the postwar era. In October 1919, the Wright-Martin Aircraft Company was dissolved, and a new company was formed: the Wright Aeronautical Corporation (Wright). The New Brunswick plant and other assets were sold, and a new factory was built in Patterson, New Jersey. Wright continued to manufacture Hispano-Suiza engines. However, Wright had made many modifications to these engines, and "Hispano-Suiza" was dropped from their names. Wright also began the development of a new water-cooled V-12 aircraft engine that was larger and more powerful than previous engines.

In 1922, the US Navy launched its first aircraft carrier (USS *Langley CV-1*) and ordered its second *(USS Lexington CV-2)* and third *(USS Saratoga CV-3)*. The navy evaluated aircraft and decided that air-cooled engines were lighter, simpler, and required less maintenance than water-cooled engines. As a result, the navy wanted leading aircraft engine manufacturers to develop air-cooled engines. However, the experimental air-cooled engines of Wright and Curtiss

Enter Curtiss and Pratt

The plight of the Liberty engine and its users was not going unnoticed by the fledgling NACA (the name changed to NASA, National Aeronautics and Space Administration, in 1959) and its prescient advisers. In 1920, the agency made the statement that the burgeoning radial-engine designs could offer far better reliability and greater weight-to-horsepower growth potential. Well, thank goodness! As a matter of course, the US Navy declared in 1921 that it would not order any aircraft that did not have air-cooled radials. Okay, all that sounds good, but where in the world would they find such new engine technology? Via contract from the navy to Charles Lawrance in 1922, an architect and aeronautical engineer, and his 1917-founded company Lawrance Aero Engines, the J-1 radial engine emerged. This, as it turns out, was a watershed moment for the world.

By 1923, as Lawrance was struggling to ramp-up his production capabilities, the Navy Department asked him to allow for the purchase of his company by Wright Aeronautical (yes, the Wright brothers), whose production prowess was a known quantity to the government. He acquiesced and his new engine marvel was enfolded into Wright's line of products and was given a name that connoted speed and movement: the J-5 Whirlwind. It produced 220 hp.

Here is a partial list of J-5 Whirlwind–powered airplanes: *The Spirit of St. Louis, Ford Trimotor, N3N trainer, Bellanca WB-2, Cessna CW-6, WACO, Stearman PT-17, Curtiss P-36, Lockheed Electra Jr., Fokker F VII,* and *Pitcairn Autogyro.* Several of these airframes were well-publicized record-breaking airplanes.

Mr. Lawrance was retained at Wright as a vice president until

Known to be exuberant, impulsive, and rather obnoxious, Glenn Curtiss was a driven spirit, a true entrepreneur. Curtiss airplanes were designed to be a cut above and quite innovative. They included their moving control surfaces, a significant improvement over the wing-warping technique developed by the Wrights. Curtiss was also quite prolific and had many varying designs emanate from his workshop and factory. (Photo Courtesy Curtiss-Wright Corporation via AAHS)

had not led to much success, and neither company was enthusiastic about pursuing development of the type.

Charles L. Lawrance, founder of the Lawrance Aero Engine Corporation, had been designing and experimenting with air-cooled engines since 1914. By 1921, Lawrance had developed a 200 hp, 9-cylinder, air-cooled engine: the J-1. The navy ordered 50 J-1 engines but felt that Lawrance did not have the capacity to fulfill such an order. The navy approached Wright and Curtiss, asking for one of the companies to merge with Lawrance to produce the J-1. Neither company was interested. The navy informed Wright that it would not purchase any engines from the company until it merged with Lawrance.

In May 1923, Wright acquired Lawrance Aero and made Charles Lawrance a vice president at Wright. Lawrance, working with engineers from Wright, continued to refine his air-cooled engine. In

1924, the latest version was the J-4 Whirlwind. The engine was accepted by the navy and entered production in 1924.

The J-4 was a 9-cylinder radial with a 4.5-inch bore and a 5.5-inch stroke. The engine displaced 787 ci, had a 5:1 compression ratio, and produced 200 hp at 2,000 rpm. The J-4 used an aluminum cylinder with a steel barrel. A cam ring was positioned at the front of the engine and geared to the crankshaft via a pinion. The cam ring rotated in the opposite direction of the crankshaft at 1/8 speed and had two tracks, each with four lobes. Via roller tappets, one track actuated pushrods for the intake valves and the other track actuated pushrods for the exhaust valves. The exposed pushrods extended from the front of the engine to the top of the cylinders, where they acted on rockers to open and close each cylinder's single intake and single exhaust valves. The J-4 was normally aspirated and did not

1925 and then, as Wright's chief decided to leave that company and start an engine company of his own, Lawrance was moved into the presidency. Thus began the Wright dynasty in piston aero engines. The eventual merger of Curtiss Aeroplane and Motor Corporation with Wright Aeronautical Corporation in 1929, thus ending a long legal feud between the two aviation pioneers, brought together 18 affiliated companies and 29 subsidiaries, which formally created the Curtiss-Wright Corporation. This constituted one of the two pillars of American aero engine design, which reshaped the aviation world in the following years. Now for the punch line.

As mentioned, Lawrance filled the position of president as it was vacated by the then-boss at Wright. That individual who departed was Frederick Rentschler. He then teamed with an existing tool-making company known by their combined name of Pratt & Whitney. I imagine that just about now the light bulbs are coming on for many of you. As we always say, "It is a small aviation world."

Completed on Christmas Eve of 1925, the Wasp (coined as its name by Mrs. Rentschler) was in response to a navy (there they are again) request for an air-cooled radial engine that would produce more than 400 hp but weigh no greater than 650 pounds. The service branch wanted the engine ready and available for 1926 flight testing on, of all things, a *Wright F3W-1!* Rentschler and his men obliged, the engine developed 450 hp, making the navy pleased, and P&W stepped squarely into the arena and the history books.

A quick look at *Wasp R-1340* (R for radial and then numeric displacement of the engine) powered aircraft includes: the *Boeing 247, Sikorsky S-38, North American T-6, Gee Bee R-1, Beechcraft D-18,* and *Lockheed Vega.* No fewer than 34,966 Wasp engines were produced over a 34-year production lifespan. Now that's significant.

George Mead, designer of the P&W Wasp stands on the far left next to Frederick Rentschler, president of Pratt. These two men did for engines what Curtiss did for airplanes in those early days and changed the world forever. Seen here at the factory in Hartford, Connecticut, the men are viewing the 1,000th Wasp engine to come off the assembly line. The relative size of then-current day technology in any aspect of aviation (other than airships) was truly not very large.

have a propeller gear reduction. The engine was 44 inches in diameter and weighed 475 pounds.

The J-4 used a one-piece crankshaft and a split, two-piece master connecting rod. The two-piece master rod was the standard practice for the era, but it caused issues in radial engines. In a radial engine, each row of cylinders is served by a single crankpin on the crankshaft. In the case of the J-4, the pistons of nine cylinders were actuated by the crankshaft's one crankpin. This was accomplished by having a master connecting rod, which served one cylinder, attached directly to the crankshaft. The pistons for the other cylinders were attached to the master rod by linked (also called articulating) connecting rods. The material available to secure the master rod to the crankshaft decreases with each additional cylinder served by the master rod. The stresses on the master rod increase as the engine produces more power. At a certain output, the two-piece master rod becomes the weak point in the engine's powertrain.

Wright engineer George Mead circumvented the master rod strength issue in 1920. Rather than using a split master rod, the master rod was a single piece, and the crankshaft bolted together through the master rod. Mead applied this configuration to the Wright R-1 engine. The engine was a 350-hp 9-cylinder radial that displaced 1,454 ci. The R-1 was the first high-power radial successfully tested in the United States. Because it was the lowest bidder, Curtiss was awarded a contract to further develop and produce the engine as the R-1454. Delays, combined with the progress of new engines, led to the abandonment of the R-1/R-1454.

Another issue with the master and link rod setup is the spacing

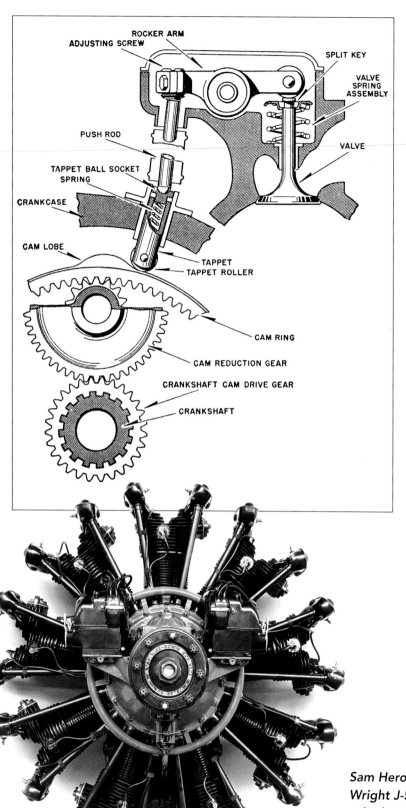

This diagram of the standard valve drive for a radial engine illustrates how the crankshaft powers a cam reduction gear, which drives the cam ring. Roller tappets ride tracks on the cam ring. Lobes on the track lift the tappet, which lifts the pushrod that actuates the valve via a rocker arm.

Designed by George Mead, the R-1 was the first radial engine built by Wright and the first high-power radial built in the United States. The engine's four valves per cylinder and general layout resembled that of the British Bristol (Cosmos) Jupiter, which was first run in 1918. (Photo Courtesy Aircraft Engine Historical Society)

Sam Heron and Edward Jones incorporated many refinements into the Wright J-5. Unlike the J-4, the J-5's valves were enclosed, and the basic cylinder shape and cylinder cooling fins exhibited a more sophisticated design. (Photo Courtesy Aircraft Engine Historical Society)

of the link rods around the master rod. Unlike with the master rod, the pivot point (knuckle pin) of the link rod is not at the center of the crankpin. Therefore, the link rods follow an elliptical path rather than a circular path as with the master rod. If the link rods are evenly spaced around the master rod, the pistons are slightly off and do not reach TDC at the perfect time.

The TDC issue can be fixed by changing the spacing of the link rods so that they are unevenly spaced around the master rod. The link rods nearest the master rod have greater spacing, and the link rods farthest from the master rod (on the opposite side) are mounted closer together. This arrangement slightly alters the piston stroke of the link rod cylinders, but that drawback is better than having the pistons' TDC timing off. Each mirrored pair of link rods have the same stroke, but the stroke is slightly different for each link rod pair and the master rod. When viewed from the rear, the cylinders of a radial were numbered clockwise, with the top cylinder as number-1. The master rod for the J-4 engine was located in cylinder number-1.

Frederick Rentschler, the president of Wright, believed that the future of aircraft powerplants lay with powerful air-cooled engines. While with the army, Rentschler was an inspector of Wright Hispano-Suiza engines during the war. After the war, he played a crucial part in the formation of the Wright Aeronautical Corporation. Rentschler realized the potential for high-power air-cooled engines and sought support from Wright's board of directors to invest more company profits into engine development. The board was not interested in providing additional funds, and Rentschler resigned out of frustration on 21 September 1924. Lawrance was appointed president after Rentschler's resignation.

Even if it was not at the pace Rentschler wanted, engine development continued at Wright. A new engine developed from the R-1 and, called the P-1 Cyclone, was first run in 1925. The P-1 engine produced 400 hp. However, Wright now had competition. By the end of 1925, Rentschler had formed the Pratt & Whitney Aircraft Company (P&W) and had a running engine more powerful than any air-cooled engine at Wright. The Wright board of directors were now compelled to invest more money into engine development, and the Cyclone was about to take off.

A further-developed version of the Whirlwind, the J-5, entered production in 1926. This engine had been refined by Sam Heron and Edward Jones, two engineers who were hired after others defected to P&W. Heron and Jones had worked to improve air-cooled cylinders while they were employed at the Army Air Corps Power Plant Lab. Their experiments led to a cylinder with a hemispherical combustion chamber and more numerous, thinner fins that increased the overall area to dissipate heat from the cylinder.

Essentially, the cylinders were formed using a steel barrel screwed

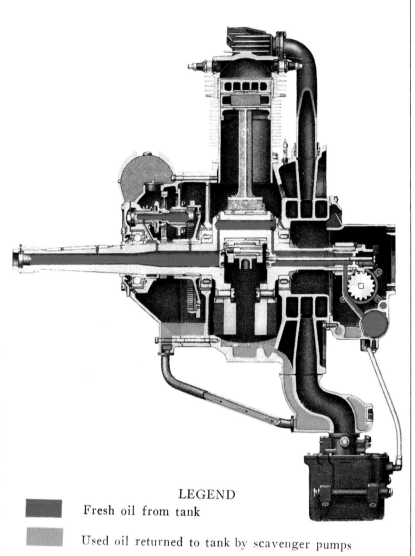

LEGEND

Fresh oil from tank

Used oil returned to tank by scavenger pumps

This sectional view of the J-5 engine and its oil system reveals a relatively simple radial engine. Gear reduction and supercharging was incorporated in later radial engines as more power was needed. Note the hemispherical combustion chamber. (Photo Courtesy Lubrication Instructions for the Wright Whirlwind Aviation Engine, July 1927)

into an aluminum head. The pushrods were enclosed in tubes that ran from the top of the cylinders to the front of the crankcase. The J-5 engine had the same bore, stroke, and configuration as the J-4, but the J-5 had a compression ratio of 5.4:1, and it produced 220 hp at 2,000 rpm. The engine was 44.25 inches in diameter and weighed 510 pounds. Heron had personally prepared the J-5 engine installed in the *Spirit of St. Louis* before Lindbergh's Atlantic flight.

In 1929, the Whirlwind was redesigned to create the J-6 series. The bore was increased to 5.0 inches, but the stroke remained unchanged at 5.5 inches. A switch was made to a one-piece master connecting rod and a two-piece crankshaft. The two crankshaft sections were joined at the crankpin by a clamping collar, with a bolt pinching the collar tight to secure the connection. The cam ring and pushrods were relocated to the rear of the engine. An induction blower (weak supercharger) at the rear of the engine helped mix the air and fuel to provide an equal distribution to all of the cylinders.

Three models of the J-6 were made that varied in their number of cylinders and outputs. All three models had a 5.1:1 compression ratio, and used 73-octane fuel, the standard at the time. The Whirlwind 165 (R-540) was the smallest. It was a 5-cylinder engine that displaced 540 ci and produced 165 hp at 2,000 rpm. It weighed 370 pounds. The Whirlwind 225 (R-760) was a 7-cylinder engine. It displaced 756 ci, produced 225 hp at 2,000 rpm, and weighed 425 pounds. The largest of the Whirlwind series was the 300 (R-975).

Wright Whirlwind Engines in Airliners				
Manufacturer	A/C Model	Engine Model	HP t/o	Number
Ford	4-AT-A	J-4	200	3
	4-AT-B	J-5	220	3
	4-AT-E	J-6-9	300	3
Lockheed	10-B	R-975E3	440	2
Sikorsky	S-36	J-5	220	2

It was a 9-cylinder engine that displaced 972 ci. The Whirlwind 300 produced 300 hp at 2,000 rpm, was 45 inches in diameter, and weighed 485 pounds.

Wright continued to develop and improve the Whirlwind engines through the 1930s and early 1940s, and power outputs rose accordingly. The R-540 maintained its 5.1:1 compression ratio, but its output climbed to 175 hp at 2,000 rpm. The compression ratio for the R-760 and R-975 was increased to 6.3:1, and the engines were rated with 80 octane fuel. The R-760 produced 350 hp at 2,400 rpm, and the R-975 had an output of 450 hp at 2,250 rpm. The R-760 and R-975 engines were produced under license by the Naval Aircraft Factory during World War II. Production totals for the Whirlwind

The Wright J-6/R-975 was the final development of the engine originally designed by Lawrance. Note the continued refinement of the engine's cylinders compared to the J-5 and earlier engines. In addition, the magnetos and valve drive were moved to the rear of the engine to provide unobstructed airflow to the cylinders. (Photo Courtesy Aircraft Engine Historical Society)

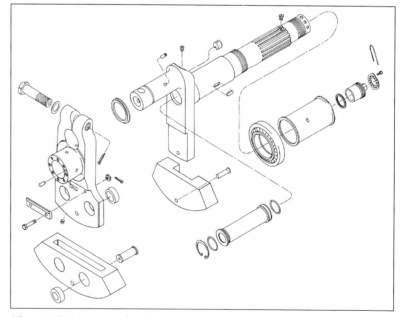

The J-6/R975 switched to a single-piece master connecting rod and a two-piece crankshaft. The built-up crankshaft used a clamping collar that slid onto an extension of the crankpin. A bolt pinched the collar tight to secure the connection. Wright used this same method of crankshaft assembly on its later piston engines. (Photo Courtesy Technical Manual Illustrated Parts Breakdown Aircraft Engine USAF Model R-975-46, January 1959)

series of engines includes 3,339 J-series, 536 R-540s, 2,541 R-760s, and 8,141 R-975s. An additional 54,103 R-975 engines were built by Continental for use in M4 Sherman and M18 Hellcat tanks.

Pratt & Whitney History Up to the Hornet B (1930)

Shortly after Frederick Rentschler left Wright Aeronautical, he decided that he wanted to continue his aircraft engine career. Rentschler sought a company that had a surplus of cash and was looking to invest in a new endeavor: designing and building aircraft engines. He found exactly what he was looking for in the Pratt & Whitney Company, a manufacturer of machine tools.

Rentschler requested $250,000 to develop the prototype of a new engine, and an additional $1,000,000 if the engine entered production. In addition, he asked that one of the tool company's mostly unused buildings serve as the offices and shop for engine design and production. In return, the new aircraft engine manufacturer would carry the Pratt & Whitney name, and the tool company would own half of the engine company and be entitled to half of its profits. However, no one outside of the engine company would have a say in how its engines were developed or its profits were spent.

The terms were agreed to, and the Pratt & Whitney Aircraft Company was founded on July 22, 1925 in Hartford, Connecticut. The navy offered its support by offering to purchase six experimental engines for $90,000.

Some Wright engineers who had expressed interest in joining any new company with which Rentschler was involved left Wright before financing had been finalized. Former Wright engineers George Mead and Andy Willgoos were joined by Earle Ryder, and the three worked on designing the new engine in Willgoos's garage, as the P&W facilities were not yet ready. The target for the new engine was an output of 400 hp and a weight of 650 pounds. While it was still in the design phase, the engine was named "Wasp" by Rentschler's wife, Faye.

The Wasp incorporated some features that Mead originally designed for the Wright R-1 in 1920. The Wasp had a 5.75-inch bore and stroke and a total displacement of 1,344 ci. Nine cylinders were spaced around a forged aluminum crankcase. A crankcase made from an aluminum forging was much stronger than one made from an aluminum casting, which was the typical construction method. The crankcase was split vertically through the cylinder. For all radials, the assembled crankcase was often called the power section.

The Wasp engines used a two-piece crankshaft and a single-piece master connecting rod. The two-piece crankshaft was split at the crankpin, and a splined male end mated with a splined female end. The two crankshaft halves were secured by a precision bolt that passed through the center of the crankpin. These design elements

The R-1340 Wasp was the first engine made by the Pratt & Whitney Aircraft Company (P&W). An absolute success, the engine was in production from 1926 to 1960. Note the thick cooling fins on the cylinders. This was typical cylinder construction for the late 1920s and early 1930s. The valve boxes are cast integral with the cylinders, a feature that originated with P&W engines. (Photo Courtesy Aircraft Engine Historical Society)

resulted in a stronger, more rigid engine that could withstand the stresses of higher power outputs. The master rod was positioned in cylinder number-1 in early engines, but it was moved to cylinder number-5 in later engines.

Each cylinder had one intake and one exhaust valve. The valve rocker boxes were cast integral with the cylinders. Pushrods concealed in tubes extended down from the rocker boxes to the front of the crankcase. Via roller tappets, the pushrods were actuated by lobes on a cam ring, which was driven at 1/8 crankshaft speed and in the opposite direction of the crankshaft. The cylinders had thinner and more numerous fins than those used on previous engines. This allowed for more cooling surface area and better heat transfer away from the engine and into the air. A rotary blower was incorporated into the rear of the engine. The blower mixed the air and fuel and provided a nearly homogeneous mixture to all nine cylinders.

Five two-piece crankshafts have been assembled through their one-piece master connecting rods and await installation in R-1340 crankcases. The head of the precision bolt used to join the crankshaft halves can be seen on the assembly at far right. Note the extraordinary finish of the components. (Photo Courtesy LOC 2017694150)

The single-piece master connecting rod was stronger than the earlier two-piece (or split) type. The master rod and crankshaft were the strongest parts of the engine. These R-1340 master rods were mass produced during World War II. (Photo Courtesy LOC 2017694103)

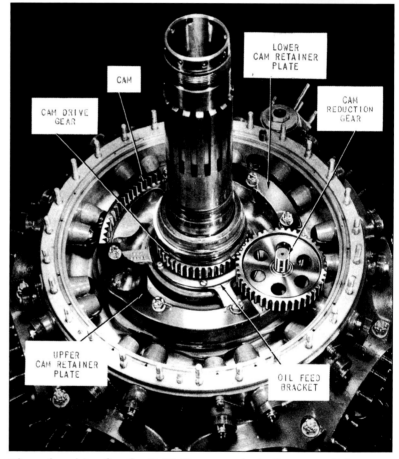

The valve drive for the R-1340 was typical of most radial engines built in the United States. The cam reduction gear was driven by the crankshaft. A smaller gear on the backside of the cam reduction gear drove the cam ring via its internal gear teeth. The lobes on the cam ring actuated the roller tappets positioned around the engine. The tappets operated the pushrods (not pictured) that led to the valves in the cylinders. (Photo Courtesy Air Depot Progressive Overhaul Manual Pratt & Whitney R-1340-AN1 Engine, December 1943)

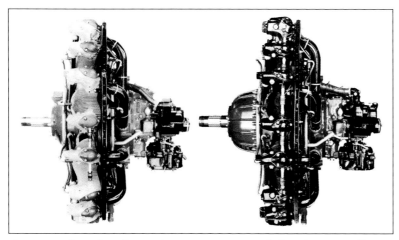

Compared to the direct-drive Wasp S3H1 (left), the propeller gear reduction on the Wasp S3H1-G (right) made the engine about 4.8 inches longer and 88 pounds heavier. The gear reduction used on the R-1340 consisted of six bevel gears. (Composite from Installation Handbook, March 1948)

R-1340 Wasp engines built during World War II had significant changes in cylinder finning and construction from earlier versions of the engine. The thinner and more numerous fins helped dissipate heat created by the higher outputs achieved with better fuels and increased supercharging. (Photo Courtesy LOC 2017694135)

Although the blower was not supercharging the incoming mixture, the possibility of using the blower to increase the induction air pressure was not overlooked by the design team.

Space was slowly being prepared for the P&W personnel, so design operations moved from Willgoos's garage to what became the P&W facility. The building had previously been used to store tobacco, and tons of it still remained. Some of the P&W employees helped "relocate" small quantities of tobacco. Colonel Edward Deeds was on P&W's board of directors. He said that he would buy everyone a turkey if the Wasp was completed before Christmas. The first Wasp was assembled on 24 December 1925; come Christmas Day, no one went hungry.

The completed Wasp weighed just under 650 pounds and had a diameter of 51.75 inches. The engine was first run on 29 December 1925 and held to 380 hp. The results from the first run showed that everything was performing exactly as it should. The Wasp was taken to 410 hp on its next run and 425 hp at 1,900 rpm on its third. By the end of March 1926, the Wasp had completed a 50-hour certification test, and five additional engines were under construction to fulfill the navy's order for six engines. At the request of the navy, the first Wasp engine never flew and was set aside for preservation. The second engine made its first flight on 5 May 1926, installed in the Wright XF3W Apache. At the time, P&W had spent around $205,000 of its initial seed money, and $1,000,000 was advanced by the tool company to enable the engine's production.

In October 1926, the navy ordered 200 Wasp engines, which established P&W as an aircraft engine manufacturer. Through continual development and the use of better fuels, the Wasp's power rating was increased to 450 hp in 1928 and 550 hp in 1933. With 87-octane

Pratt & Whitney R-1340 Engines in Airliners				
Manufacturer	A/C Model	Engine Model	HP t/o	Number
Bach	3-CT-9	SC-1	450	1*
Boeing	80	Wasp A	410	3
	247	S1D1	550	2
	247D	S1H1-G	600	2
Fokker	F.10	Wasp A	410	3
	F.10A	Wasp B	450	3
Ford	5-AT-A, B, C	C-1	420	3
	5-AT-D	SC-1	450	3
Lockheed	10-C	SC-1	450	2
Sikorsky	S-38	SC-1	450	2
* The Bach Air Yatch 3-CT-9 had two additional smaller engines.				

fuel, the engine reached a takeoff rating of 600 hp at 2,250 rpm and a normal rating of 550 hp at 2,200 rpm. The Wasp engine's compression ratio was initially 5.25:1, but it was increased to 6.0:1 over the life of the engine. Later versions of the engine were available with a .667 propeller gear reduction achieved through planetary gears.

Ultimately, 34,966 R-1340 Wasp engines were made, many of which were built under license during World War II. The engine was in production for 35 years, until 1960.

As if starting a new company and building a new aircraft engine were not enough, Mead and Willgoos had been hard at work designing a larger engine; the Hornet. It was essentially a scaled-up version of the Wasp, and 87 percent of its parts were interchangeable with those used on the Wasp. The engine had a 6.125-inch bore and a

6.375-inch stroke. Total displacement was 1,691 ci, and the Hornet produced 525 hp at 1,900 rpm with a 5.0:1 compression ratio. The engine was 54.5 inches in diameter and weighed 750 pounds.

The first Hornet was finished in June 1926, and its 50-hour certification test was successfully completed on 25 March 1927. In 1933, the Hornet was rated at 700 hp at 2,050 rpm. With its compression ratio increased to 6.5:1 and using 87 octane fuel, the engine achieved a takeoff rating of 875 hp at 2,300 rpm and a normal rating of 750 hp at 2,250 rpm. The later, more powerful versions of the Hornet were available with a .667 planetary gear reduction for the propeller. The R-1690 Hornet was in production until 1942, and a total of 2,944 engines were made.

In 1928, an enlarged version of the Hornet was designed. Known as Hornet B or Big Hornet, the engine first ran in 1929. The Hornet

In many regards, the P&W R-1690 Hornet was an enlarged Wasp. The Hornet powered the Boeing Model 40B single-engine mail plane. Some Boeing 40Bs were converted to carry passengers, and this was how the air carriers that became United Air Lines got their start. This is a Hornet equipped with a propeller gear reduction. (Photo Courtesy Aircraft Engine Historical Society)

This rear view of the Hornet displays individual induction manifolds that took the air and fuel mixture from the supercharger and delivered it to the cylinders. Engine mounting pads adjacent to the induction pipes can be seen on the supercharger housing. (Photo Courtesy Aircraft Engine Historical Society)

B had a 6.25-inch bore, 6.75-inch stroke, and total displacement of 1,864 ci. The engine had a 5.0:1 compression ratio and produced 575 hp at 1,800 rpm. Its diameter was 57 inches, and the engine weighed 985 pounds. For the time, the Hornet B was hitting the upper limit of what a single-row 9-cylinder radial could displace and produce. Although its output rose to a takeoff rating of 650 hp at 2,000 rpm and a normal rating of 600 hp at 2,000 rpm, P&W was never completely satisfied with the engine. It is interesting to note that each of the Hornet B's cylinders displaced 207 ci and that no successful air-cooled aircraft engine has employed cylinders of equal or greater displacement. Only 446 R-1860 Hornet B engines were made.

By 1929, the rapid rise of P&W saw the company outgrow its building, and a new plant was built at East Hartford, Connecticut, with 400,000 square feet of manufacturing space and 30 engine test houses. P&W bought out the tool company's share of ownership. Rentschler worked with William Boeing, Chance Vought, and Thomas Hamilton to create an aviation conglomerate. The result was the merger of P&W, Boeing Airplane Company, Chance Vought Aircraft Company, and Hamilton Aircraft Company to form the United Aircraft and Transport Company. Sikorsky Aircraft Corporation, Stearman Aircraft Corporation, and Standard Steel Propellers joined the conglomerate a short time later. In 1930, Rentschler left P&W to head its new parent company, and Don Brown took over as president.

The quick success of the powerful and reliable air-cooled engine resulted in the navy's abandonment of water-cooled types. Although the navy occasionally funded development of water-cooled engines, only air-cooled types were selected for service.

Lycoming R-680

Wright Aeronautical and P&W were not the only manufacturers in the United States building air-cooled radial engines for commercial use. Lycoming also provided engines to power Stinson's line of trimotor airliners. The Lycoming Foundry and Machine Company

The P&W R-1860 Hornet B was essentially an enlarged Hornet A, which itself was an enlarged Wasp. However, the 6.25-inch bore and 6.75-inch stroke of the Hornet B's 207-ci cylinder proved to be a bit too large for P&W's technology in the early 1930s. Only a small number of the engines were built, and P&W used smaller cylinders on all of its subsequent piston engines. (Photo Courtesy Aircraft Engine Historical Society)

Pratt & Whitney R-1690 Engines in Airliners				
Manufacturer	A/C Model	Engine Model	HP t/o	Number
Bach	3-CT-6, 8	Hornet A	525	1*
Boeing	80A	Hornet A	525	3
Lockheed	14-H	S1E-G	875	2
	14-H2	S1E2-G	875	2
	18-07	S1E2-G	875	2
Sikorsky	S-40A	T2D1	660	4
	S-42	S1E-G	875	4
	S-43	S1E-G	875	2
* The Bach Air Yatch 3-CT-6 and -8 had two additional smaller engines.				

Pratt & Whitney R-1860 Engines in Airliners				
Manufacturer	A/C Model	Engine Model	HP t/o	Number
Fokker	F.32	Hornet B1-G	575	4
Sikorsky	S-40	Hornet B	575	4
	S-41	Hornet B1	575	2

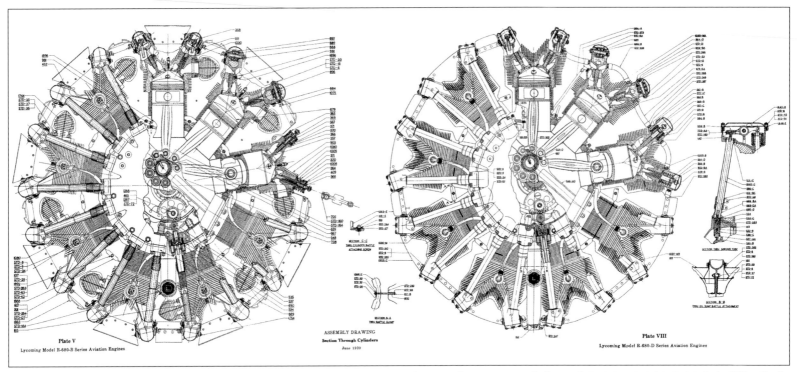

Plate V
Lycoming Model R-680-B Series Aviation Engines

ASSEMBLY DRAWING
Section Through Cylinders
June 1939

Plate VIII
Lycoming Model R-680-D Series Aviation Engines

The 240 hp R-680 B (left) and 260 hp R-680 D (right) differ little except for cylinder development. To dissipate heat as more power was produced in the cylinders, cylinder cooling fins were developed to be progressively finer, deeper, and more numerous. (Composite from Operators Manual: Lycoming R-680 and R-530 Aviation Engines, *June 1939)*

was created in 1907 at Williamsport (Lycoming County), Pennsylvania. By 1910, the company was focused on manufacturing automotive engines that were used in Auburn, Checker, Cord, Duesenberg, Graham, McFarlan, Paige, and Velie automobiles. Auburn, Cord, and Duesenberg were owned by Errett Lobban Cord, and Cord bought Lycoming in 1927.

In the late 1920s, Lycoming built its first aircraft engine, a 9-cylinder air-cooled radial. The engine had a 4.5-inch bore and stroke and displaced 644 ci. Known as the R-645, the engine was built like contemporary radials, with a cast-aluminum crankcase and cylinders constructed from forged steel barrels that were screwed

The Lycoming R-680 was a reliable engine and powered a number of Stinson airliners. The engine choice is not surprising considering that Lycoming and Stinson were both owned by the same company. The R-680's most successful application was in the Stearman PT-13 biplane trainer, of which more than 2,100 were made. Note the finned exhaust ports on this Lycoming R-680 B. (Photo Courtesy Aircraft Engine Historical Society)

First flown in May 1934, a year after the Boeing 247 and a month before the Douglas DC-2, the Stinson Model A was the last trimotor airliner built in the United States until the Boeing 727. It was also the last airliner powered by Lycoming R-680 engines. The humps on the cowlings provided clearance for the engine's valve rockers. Note the wing-to-upper-fuselage strut. (Photo Courtesy John Oxley Library, State Library of Queensland, image number 189637)

and shrunk into cast-aluminum heads. The R-645 never entered production.

In 1929, the R-680 was created by increasing the bore of the R-645 engine. The R-680 had a cylinder bore of 4.625 inches, an increase of 0.125 inch from the R-645 engine. The stroke remained unchanged at 4.5 inches, and the engine displaced 680 ci. Each of the R-680's cylinders had one intake valve and one sodium-cooled exhaust valve. The cam ring was positioned at the front of the engine and turned in the opposite direction as the crankshaft at 1/8 crankshaft speed. The cam ring had two tracks, each with four lobes.

The R-680 engine used a one-piece master rod and a two-piece crankshaft, which was secured by a clamping bolt as on the Wright crankshafts. The master rod was installed in cylinder number-7. The crankcase was a one-piece aluminum casting to which the rear accessory housing and two-piece front section were attached. These housings were initially made from cast-aluminum but were made from magnesium for later engines. The R-680 was direct drive and incorporated a blower (weak supercharger) to mix and evenly distribute air and fuel to the cylinders.

With a compression ratio of 5.3:1, the R-680 initially produced 215 hp at 2,000 rpm. The engine had a 43.5-inch diameter and weighed 515 pounds. The R-680 was first flown on 3 April 1929 in

a Travel Air Model L and entered production later that year. The R-680 B-series was developed around 1931 and had a compression ratio of 6.5:1. The R-680 B was rated for 240 hp at 2,000 rpm. The D-series followed in 1933 and produced 260 hp at 2,300 rpm for take-off and 245 hp at 2,100 rpm for normal operation. The final development of the R-680 was the E-series. These engines had a compression ratio of 7.0:1 and an output of 300 hp at 2,300 rpm for takeoff and 285 hp at 2,200 rpm for normal operation.

The R-680 proved to be a reliable engine, and Lycoming claimed the engine was referred to as "the old sewing machine" because it "goes and goes." The nickname could also have something to do with the fact that Lycoming's predecessor built sewing machines. Stinson continued to use the R-680 through the mid-1930s, but Lycoming did not pursue another radial for commercial aviation. The engine was popular in smaller aircraft and trainers, and a 7-cylinder version of the engine was built: the R-530. Around 25,000 R-680s were built, and production continued through World War II.

Stinson's trimotor aircraft were the only airliners powered by the Lycoming R-680 engine. Cord purchased Stinson in 1929, and it made business sense for engines produced by the Cord empire to power aircraft produced by the Cord empire. Stinson airliner production began in 1930 with the SM-6000, and 11 aircraft were built. The SM-6000B (Model T) followed in 1931, and it proved to be the most popular of the Stinson trimotors, with 42 examples produced. A new design, the Model U, started production in mid-1932, and 24 of the type were made. The final Stinson airliner was the Model A. The Model A made its debut in late 1933, and its production run totaled 31 aircraft.

In 1933, Lycoming built its first horizontally opposed engine. The company became part of the Aviation Corporation (AVCO) in 1939. Today, Lycoming continues to provide a variety of air-cooled, horizontal engines that power more than half of the world's general aviation aircraft.

Lycoming R-680 Engines in Airliners				
Manufacturer	A/C Model	Engine Model	HP t/o	Number
Stinson	SM-6000	R-680	215	3
	SM-6000B / Model T	R-680	215	3
	Model U	R-680-BA	240	3
	Model A	R-680-D5	260	3

Chapter 2

ONE WING, TWO ENGINES, ALL METAL

As the fourth decade of the 20th century dawned, the stock market had crashed, and the haunting specter of economic depression loomed ever larger. This environment lent very definite coloration to the 1930s period of aviation. It was a time of surprising advancement as well as deep adversity. An old saying comes to mind: When the times get tough, the tough get going. Men and women of aviation were indeed tough.

An actual color photo of an American DST at Newark airport in 1937. These were the days when airlines often built their own terminal buildings as can be seen by the TWA rotunda in the background. (Photo Courtesy Boeing Co. via Jon Proctor with permission)

Getting There by Air

In 1936

Three signal and pivotal events unfolded that cast the die for decades to follow and certainly for the immediate and ensuing 10 years. These marquees were: the airmail acts of 1930 and 1934, the Boeing 247, and 31 March 1931. Each spawned immense, world-altering changes within airline operations.

The year 1930 marked the McNary-Watres Act, allowing the postmaster to choose, as per his own judgment, the airlines to be awarded the lucrative airmail contracts from the government. No more competitive bidding among companies, which also favored the largest of the airlines and drove small carriers out of business.

With the incoming Roosevelt administration setting the tone for congressional Democrats, a hearing "determined" that collusion had existed between Walter Folger Brown and his favored airline companies in assigning the awards. This led the president to cancel all existing contracts via the Air Mail Act of 1934. Overnight, the airlines were stripped of mail revenue, and the government was once

The Stinson Model T or SM-6000. E. L. Cord was a part owner in American Airways as well as Stinson and so naturally, the airline used the latter's product. This airplane joined the fleet in 1932 and carried 10 passengers at a whopping 125 mph. Both National and Delta were among the several domestic carriers to operate the Model T.

The Stinson Model U was a refinement in both speed and quality from the Model T. It carried the same 10 fares but in slightly larger and nicer accommodations. In 1932, this airplane new from the factory cost $19,500 each. American was the only carrier to operate the airplane in large quantity, while both Eastern and National dabbled with one each.

On the small but jaunty side of the ledger is the 1930 Stinson Model A. Rakish and low-wing, the A carried eight passengers. Thirty airplanes were produced before this design was eclipsed by the all-metal Boeing 247 and Douglas DC-2. (Photo Courtesy Tim Williams via Jon Proctor)

Being a passenger in the 1920s and 1930s meant being pampered by the airline, which took their riders very seriously because there were so few of them! After buying your ticket at a downtown hotel, such as the Parker House or Fairmont in Boston, the Colonial Airlines limo whisked you to the East Boston airport for your flight to New York.

A nice study of 1930s chic: A Douglas Dolphin amphibian moored in Wilmington, California, not far from the Big White Steamship, which ran between Catalina Island and the mainland. (Photo Courtesy Mike Machat Collection)

Chapter 1 included a Curtiss Condor 1 airliner with its inline engines. Here is its interior configuration. There was an obvious late-1920s aesthetic and design sense to the airplanes starting with the Ford and progressing through the Stinsons and Curtiss Condor series. (Photo Courtesy Gerald Balzer Collection)

It took a great many hours to cross a country the size of America traveling at 100-plus mph. The solution for increasingly longer-duration flights was a sleeping arrangement of some kind. Enter the Curtiss Condor 2 and its expanded fuselage width. Now there was room to erect berths similar to those used in train travel, allowing for a more refreshing flying experience.

This cutaway illustration highlights a Condor 2 configured for American Airlines in 1933. As luxurious as it might appear, the airplane was built around welded steel tubing with fabric covering and was therefore fairly noisy. (Photo Courtesy Gerald Balzer Collection)

Swissair was another T-32 Condor customer. Aerodynamic considerations required the upper wing to be separate from the fuselage, which took the structure to soaring heights. (Photo Courtesy Gerald Balzer Collection)

Reminiscent of the Ford cockpit in chapter 1, the Spartan nature of the Condor's nerve center is remarkable for the early 1930s. (Photo Courtesy Gerald Balzer Collection)

again charged with flying the mail. This had disastrous results in its implementation and operation.

The army was completely unequipped to perform this mission, both from aircraft and training/experience aspects. The experiment lasted only three months when a new Air Mail Act in June 1934 reissued commercial carrier mail contracts. The "close-call" nature of this ill-conceived sojourn was not lost on airline managements across the country and shaped the criteria eventuating the world's most incredible airplane of its time, some say even to this day: the Douglas DC-3.

Before all the airmail hubbub, aviation pioneers were acting on their visions to advance aeronautics for airline passenger travel. The designers in Seattle, Washington, took the early lead right out of the gate with a revolutionary airplane known as the Model 247. This single airplane was the gestating point of the truly modern airliner, and the world took due notice. Here was a sleek, all-metal monoplane, twin-engine (new P&W Twin Wasp Juniors of 625 hp each) airliner that could carry 10 fares at 198 mph for 650 miles. That was faster by a large amount over the Fords and Fokkers of the time and could climb, fully loaded, on only one of two operating engines.

This was a remarkable breakthrough airplane. So dramatic was all this technology that several airlines clamored for rights to purchase 247s. However, as noted in chapter 1, Boeing Air Transport (later United Air Lines) was an in-house entity that had "first dibs" on anything Boeing. It took advantage of its position and ordered 70 airframes, all the first 70 off the production line, which seemed at the time a masterstroke of competitive élan, having the effect of locking-out the competing airlines across the country. United was in the catbird seat.

Just prior to the unveiling of Boeing's wonder-liner, however, one spring day in 1931 will forever be remembered as the starting date of the financially successful airline business in the United States. On 31 March, a Transcontinental and Western (TWA) Fokker trimotor found itself engulfed by a Midwestern thunderstorm over Kansas. The storm tore apart the wooden wing of the airliner and all souls aboard were lost. One of the passengers was the famous Knut Rockne, legendary head football coach at Notre Dame University.

The death of this giant man shocked the nation, a remorse felt even deeper when it was determined that the wing of the Fokker, made of wood, had actually rotted through. It was as easy as snapping a matchstick in the vicious cross currents of the storm cloud. Wood-winged airliners with their multi-ply bonded spars were impossible to inspect for this deterioration and were basically cast aside from that point forward. This one revelatory air mishap put in motion the search for and development of all-metal airframes, thus the 247. And it didn't stop with Boeing.

Over at TWA, President Jack Frye found himself in the direct line of fire due to the Rockne crash. As he looked beyond the immediate need to keep the airline running without its Fokkers, new technology was the answer to the problem of replacing the old-line airplanes. His initial quest, however, was stymied by United having taken and Boeing allowing the blocking of the first year of 247 production. He would not wait in line for slots well after United was flying its pants off, beating TWA.

The curtain opened on the future of airline flight when Boeing created the 247. Often noted as the first "modern airliner" in US fleets, the twin monoplane was a thing of aerodynamic beauty. The all-metal aircraft was a milestone in the progression to advanced airline flying; just compare this shape to the Condor. (Photo Courtesy Jon Proctor)

There were some shortcomings to the 247; however, most notable was the main wing spar that traversed the fuselage. Causing passengers to climb over the wing box to take their seats in the forward fuselage was an inspiration to Art Raymond and Donald Douglas in Southern California. They designed the DC-1 specifically to avoid this inconvenience and thus, also grew the size of the fuselage. (Photo Courtesy Gerald Balzer Collection)

In California, Lockheed had reconstituted itself and was manufacturing small single-engine airliners. This is the retractable-gear Orion 9D, a fast airplane that found wide use by individuals and airlines alike. The company liked using wood and metal in combination for its structures. The fact that this is a monoplane was a harbinger of things to come. (Photo Courtesy Gerald Balzer Collection)

Also from Southern California was Jerry Vultee's V-1 airliner. It was a fabulous airplane, but alas, the government edict that no air-liner may be powered by a single engine effectively killed this brilliant design. This particular aircraft had been bailed by Shell Oil Company for use by its ace pilot and vice president, Jimmy Doolittle, who flew the airplane cross-country in a promotional tour. The V-1A for American Airlines had a single 735-hp Wright Cyclone engine.

The world's most-famous request for proposal was then written by Frye. A letter with his specifications was sent on 2 August 1932 to four American airframe manufacturers: Curtiss, Consolidated, Douglas, and Martin. The airplane required was to be a trimotor of all-metal construction with the latest navigation and aeronautical design features, a range of 1,000 miles, cruising altitude of 21,000 feet, 14,200 pounds gross weight, and at least 146-mph cruising speed; a tall order from the man in Kansas City.

A second aero design renaissance was taking place in Southern California at the very moment Frye's letter hit the desk of Donald Douglas. One of his promising engineers was a young Jack Northrop. Northrop's design genius birthed innovative wing designs, and the Douglas proposal received a multi-cellular wing. This allowed the wing spars to run underneath the cabin versus Boeing's through the 247 cabin design. This allowed the cabin to be taller and obstruction free and cast the mold for airliners from that moment forward. But how to achieve the performance guarantee? Both Curtiss and Pratt offered much-improved engines, which made the next quantum leap in airliner design a reality.

For Curtiss it was the SGR-1820-F Cyclone, and Pratt gave up the Twin Wasp. This leap in horsepower allowed Douglas Chief Designer Art Raymond to convince a skeptical Charles Lindbergh and Jack Frye that two rather than three engines would work just as well and cost less money to run and to maintain versus the trimotor concept Lindbergh had insisted on with Frye. Lindbergh therefore had to insist that the airplane be able to take off with a full payload from any TWA station but do so with one engine out! Douglas took the requirement and agreed to it, and the team won the contract. The resulting airplane was the DC-1.

The first of the Douglas Commercial family made its maiden flight in July 1933 and actually lived life as a one-off prototype/proof of concept airplane. After TWA worked with the airplane for a few months in real-world operations, the improved production version was approved by Frye and company and was known as the DC-2. Douglas and the airlines were off to the races.

The DC-2 was a worldwide success from China to Switzerland and everywhere in the United States (except of course, at a very cha-grined United). It carried 14 passengers and was dramatic to be in

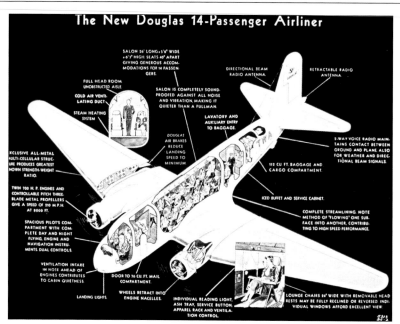

The next logical step for the world of air travel was the Douglas DC-1 (Douglas Commercial number one). This is the airplane that Jack Frye wanted, and it did not disappoint.

The DC-1 was a "flying laboratory" prototype airliner. The production version was actually the DC-2. Several refinements and a one-window fuselage stretch, and the airplane was ready for revenue service. This is a company cutaway highlighting everything from the capacious fuselage cabin to steam heating to reclining lounge chairs; all the comforts of home! (Photo Courtesy Mike Machat Collection)

TWA flew the airplane in all types of route-proving trials, and here is the ship sitting at Grand Central Air Terminal on its way east to New York via Kansas City. Fans of Buck Rogers will admit that this revolutionary airplane design might as well have come from outer space. Remember, this was mid-1933, a scant 15 years after the Great War with its strut-and-wire biplanes and coincident with the last gasp of the old days, the Curtiss Condor. (Photo Courtesy Museum of Flying Collection)

A DC-2 on the outdoor finishing ramp at Santa Monica. This airplane was part of an Eastern Air Lines order and provided quality jobs during the Great Depression. (Photo Courtesy Museum of Flying Collection)

Narrow but high is the best way to describe the DC-2 cabin. TWA had already installed club seating on the starboard rear side of the cabin, a design layout that is still popular today. (Photo Courtesy Museum of Flying Collection)

American Airlines and the Douglas Aircraft Company, recognizing the potential of film for promoting air travel, cooperated in the production and distribution of Fox Pictures' Bright Eyes in 1934. Douglas provided a real DC-2, designated "A-74," for the exterior shots, and a full-scale cabin mockup was provided for the interior scenes. Pictured here are five-year-old child actress Shirley Temple and actor James Dunn. (Photo Courtesy Mike Machat Collection)

This was the only flying DC-2 in the world in 2014. Restored by volunteer workers at Douglas, this airplane is now on display at the Museum of Flight in Seattle, Washington, at Boeing Field; irony notwithstanding. I have been aboard this airplane and at 6-feet-1-inch, I can assure you that the "New giant Flagship" of old is pretty darned small. (Photo Courtesy Jon Proctor)

The famous Eastern Air Lines steward brigade. White mess dress jacket and gloves made these men a cut above the normal nascent cabin staff of other lines.

The man with the vision, in 1934 and for many years to follow: C. R. Smith, CEO of American Airlines.

and around. However, the airlines still needed mail contract subsidy to make money at its businesses. Hauling only paying passengers was not enough. This was about to change thanks to the inspired epiphany of another airline boss of gigantic proportions: Cyrus Rowlett Smith of American Airlines.

The World's Most Influential Airliner

Working with his chief engineer (all airlines had engineering departments in those days and for years to follow), Bill Littlewood Smith devised a larger and nearly all-new airplane that would accommodate 14 passengers in a sleeper configuration, or a whopping *21 fares* in the day-plane layout of seats only. This airplane, as conceived, would finally free American from the bondage to government mail subsidies and allow it to fulfill its intended mission of flying passengers only, if desired, in order to make a profit. Now Smith had to sell Douglas on actually building the thing.

As with all the other firsts of the new Douglas, the "sales" call from Smith to Donald Douglas was also remarkable. It lasted just over two hours by long-distance toll from New York to Los Angeles, and because calling was charged by the minute in 1934 the grand total

The pioneer who acquiesced to Smith's vision, Donald W. Douglas, stands in front of a newly minted DC-3 with his equally new Buick. (Photo Courtesy Museum of Flying Collection)

Ergonomics enters cockpit design. Note the soundproofing/insulating blankets on the sidewalls and the window defrosting air blowers at each main window. This last innovation was worth its weight in gold, as in Stinson days, the copilot had to open his side window, and lean forward and out to physically scrape the ice from both windshields. Seniority has its perks. (Photo Courtesy Museum of Flying Collection)

cost for this unusual engineering communication was a not unsubstantial $335.50. If this were converted to today's inflated dollars it would amount to somewhere in the vicinity of $7,000, a small price to pay for making aviation history! American promised to buy 20 airplanes, half of them sleeper transports, and Douglas agreed to the venture. The DC-3 was born, and the world, and American Airlines, would never be the same.

It was 17 December 1935 that became an auspicious day for the DC-3 as it made a harrowing first flight from Santa Monica, California. It almost crashed several times due to incorrectly installed engine carburetors on its Wright 1820s. (Horsepower on the latest Wright engines for the DST/DC-3 ranged from 1,000 to eventually 1,200, again making possible the incremental increase in size and weight of the modern airliner airframe.) Ills were eventually corrected, and the first Douglas Sleeper Transport entered service with American as *Flagship Illinois* on 25 June 1936 from Chicago to New York and vice versa.

The seismic shift in air travel had occurred, and American was the first to enjoy it. All other airlines had to follow suit. Douglas became "airliner provider to the world."

The experience of crossing our great expanse of country in the mid-1930s was imbued with the best First-Class service an airline could provide. American had its airplanes paneled with rich woods, plush seats, curtains on the windows, and club-style seating arrangements. Passengers were separated by a large folding table if so chosen, upon which they could write letters (on airline stationery), play cards, follow their route on an airline-provided map, or allow children to play games and color pictures. Of course, the table was set with linen and china and a flower arrangement at dinner time. Standard American Airlines hot meals included fried chicken and mashed potatoes, salad, and a desert. The carrier had devised giant thermoses to hold the hot and cold foods in the galley, another first aboard this radical airliner.

Flying end to end the broad distance of the United States in an airliner that cruised at a leisurely speed of 192 mph explained the need for sleeping accommodations aboard the DST. Hinged, stowable beds in the cabin sidewalls, similar to Pullman berths in a railway car, were deployed by the single stewardess as night closed in. The daytime seats and eating tables were converted to beds with a blanket and pillows, and the 14 passengers drew curtains and got some shuteye as the airplane made its way across the country in the

There was nothing like it for domestic air travel in the United States. C. R. Smith was enamored of naval pomp and circumstance and thus named all his American Airlines fleet as Flagships, naming each after a state or a city. The red carpet certainly was out for passengers on one of American's DSTs plying the airways in both directions, east and west. Let's take a typical flight aboard one of these luxury sky ships in the year 1936.

Pages of an enticement to retailers to display a working cutaway model of the phenomenal DST Flagship. The model was valued at $1,000, which in the mid-1930s was a fortune. This type of public relations outreach indicates that the DST was indeed a very big deal. And just imagine flying aboard it all the way to California.

Much of the time, the flight schedule required flying during the night, but there were beds aloft to nestle into as you wing your way across the southern geography of America.

A Douglas Sleeper Transport at Newark, New Jersey. (Photo Courtesy Boeing Co. via Jon Proctor with permission)

Your journey begins with a limo ride to the airfield. Here is the 1938 Packard used by American to shuttle its downtown passengers.

Your beautiful Flagship, with its cabin being cooled by a portable air-conditioning truck, awaits your entry into its magic carpet world.

Welcome aboard your Flagship! We are so pleased to have you with us tonight.

Your flight crew is ready, with nerves of steel and keen senses, to guide your skyliner safely and competently to its destination.

Colorful pages highlight the airway and its attractions en route to your sunny vacation far away from the bitter Northeast winter.

An intermediate stop in El Paso to refuel, the sun is bright and warm in southern locales. This is a rare Pratt-powered DC-3 day plane with the entry door on the airplane's left side.

Meal service on a table, and with china! Like a miniature cruise ship, your DST will host you in the lap of luxury.

August 1938, Chicago; American flies in and displays a DC-3, "Flagship Chicago" appropriately enough, at Congress Street Plaza on Chicago's Michigan Avenue as part of the city's Aviation Week. Similar exhibitions of DC-3s and DSTs took place in New York and Boston. (Photo Courtesy Lamont DuPont Jr. via Jon Proctor)

Just a few of the table setting items used aboard the Flagship for mealtime. Note the specially designed napkins and silverware. The handles of the flatware have a replica of your Flagship (complete with lettering!) molded in, so every bite you take reminds you of your superb treatment aboard American Airlines.

dark. The routing for American was as originally approved by Washington and, due to the range of the airplane, necessitated several stops during the flight.

The inauguration of transcontinental flying for the DST was on 18 September 1936 and consisted of flight segments of Newark-Memphis-Fort Worth-El Paso-Phoenix-Glendale. Passengers remained in their beds for landings and takeoffs.

Airplanes were unpressurized at that time, which forced low cruising altitudes of between 10,000 and 12,000 feet. The bad news was that trips were always flown in any en route patches of weather; the good news was the view with the Douglas picture windows provided expansive landscapes for eager passengers with cameras. All in all, flying was a comfortable and mostly positive adventure for those who could afford the fare of $160 each way (when converted to 21st-century money, it would be about $3,200).

American Airlines operated the Douglas twins from 1936 to 1949 and enjoyed a total of 114 airframes in its fleet. What the DC-3 did for passenger carriage was absolutely astounding and created the modern airline business. From 462,000 airline passengers in 1934 to the 1940 tally, the number increased nearly *ninefold,* to more than 4 million.

Altogether, Santa Monica's factory produced 349 airplanes for passenger use, and after World War II, military versions boosted the final count of the DC-3 type airplane manufactured nationwide to 10,655 airframes. It was noted by General Eisenhower as one of the five most important pieces of materiel credited with winning the war. All of the many commendable results yielded by this single airplane type are not too shabby for the investment of $300 by C. R. Smith!

Serenely flying the Midwestern airways, this DC-3 during the war was one of the few at TWA with the passenger door on the right side of the fuselage. (Photo Courtesy Jon Proctor)

Airplanes always attracted attention in the early days and these folks at TWA's main hub at Kansas City were no exception. (Photo Courtesy Jon Proctor)

American wasn't the only airline to use the DST on its routes. TWA's Sky Sleeper fleet featured the near-identical furnishings for its transcontinental passengers. (Photo Courtesy Jon Proctor)

A typical nighttime launch of a *Sky Sleeper* heading west with its charges. (Photo Courtesy Jon Proctor)

Even though this is a staged ad, there is much to see and learn from this bit of local color. Shot at Burbank's Union air terminal, especially interesting is the short covered causeway at gate five. This photo layout must have been shot just prior to the outbreak of hostilities for World War II.

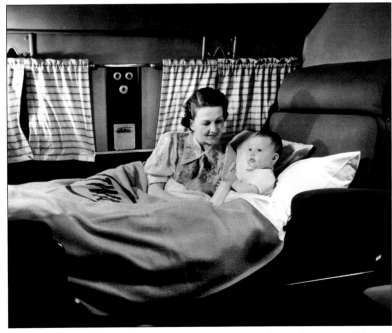

Here is a peek at seats turned into beds onboard a DST. (Photo Courtesy Jon Proctor)

The launch customer for the Boeing 247, United had to capitulate and buy DC-3s/DSTs to remain competitive. The galley for the Douglas was always at the rear of the cabin, and this stewardess is busy prepping meals.

And now dinner is served! Note that table linens are still a staple of inflight fare today. The airplane could carry enough food to serve its 21 passengers.

Almost old hat was an airline trip with a berth and breakfast in bed. The airline world was becoming more civilized by the day. (Photo Courtesy Museum of Flying Collection)

One of United's fleet emerges from the factory at Santa Monica on 8 November 1936. This is a DC-3A with P&W engines. Note the DC-2 in the immediate background and the older-style propellers and hubs. (Photo Courtesy Museum of Flying Collection)

This is how the original terminal at San Francisco's Mills Field looked, both when new and vacant and then filled with customers. The architecture is stunning. (Photo Courtesy Collection of SFO Museum)

United's main terminus was San Francisco with one of its DC-3As in front of the terminal. Note the use of a conveyor belt to hoist luggage and mail into the high front hatch. Note also the unusual addition of prop spinners on this airplane and the first use of Mainliner titles. (Photo Courtesy Collection of SFO Museum)

Looking for all the world like troublemakers, these bankers are on their way to a convention of some sort via an American Airlines DC-3 in 1938. We can see no fewer than three placements of the word "American" on the airplane and ramp, leaving no doubt about the carrier chosen for this flight.

We would be remiss to not mention the important niche filled by the up-and-coming California manufacturer that would later be so absolutely pivotal in airliner development. When the Vultee V-1 was too small (plus it had only the one CAA frowned-upon engine) and the DC-3 too large, what was a struggling airline in 1930s' depression era America to do? Why, choose a Lockheed twin, of course.

The company had a full line of three airplanes of incremental size, each developed in succession, to fit the growing needs of a young carrier, or perform at the small-end portion of a given market on the mainline. Nearly every airline in the country found a use for the Lockheed product, from Pan American Airways (PAA) to Delta, United to National. International sales were also encouraging, especially with the Lodestar. The airplanes were speedy, reliable, and rugged. They bore the influence of the renowned and soon legendary designer, Kelly Johnson, his favorite touch being the trademark twin vertical stabilizers found on each airplane.

The refined and stately Lockheed 18 Lodestar flying for Alaska Star Airlines. (Photo Courtesy Gerald Balzer Collection)

Northwest and Delta were fervent users of the small Lockheed airliners, such as this Electra 10A. One engine is all it took to keep this airplane airborne. (Photo Courtesy Gerald Balzer Collection)

Chicago & Southern with its Mississippi Valley routes didn't need a large airplane, and so the Electra was a perfect fit. (Photo Courtesy Gerald Balzer Collection)

Not able to fit into a viable revenue niche, the Model 14 was neither fish nor fowl and didn't enjoy much of an airline career but did a bang-up job as the Hudson bomber for the RAF. (Photo Courtesy Gerald Balzer Collection)

Airplane Type	Passengers	Speed	Engine Type
L10 Electra 1935	10	190 mph	PW Wasp Jr. 450 hp
L14 Super Electra 1937	14	230 mph	PW Hornet 950 hp
L18 Lodestar 1939	18	260 mph	PW Wasp/Wright Cyclone 875–1,200 hp

For route maintenance or airline business errands within the system, the "executive" airplane was the smaller Lockheed 12 Electra Junior. (Photo Courtesy Gerald Balzer Collection)

It was finally the Model 18 that made real inroads into Douglas territory by hauling up to 18 folks over some intermediate route lengths. Even PAA found the airplane effective.

Talk Lockheed with an Alaskan pilot...

In the Arctic you'll find Lockheeds performing some of aviation's toughest jobs.

Pacific Alaskan Lodestars hold to split-minute schedules in Alaska... Juneau to Fairbanks to Nome, with passengers snug at 70° inside, while the thermometer hits 30° below outside.

From Amazon Jungle Damp to Alaska, you'll find Lockheeds the stand-by of airlines that know the day-in-and-day-out grind of commercial flying.

Then too, you find these same airplanes, adapted as bombers in the heat of battle in Europe—sticking it—taking it—and bringing their crews home safely.

Lockheeds are international citizens.

... for Protection today and Progress tomorrow

LOOK TO *Lockheed* FOR LEADERSHIP

A luxurious Pan-American Lodestar flying over typical terrain of the far north along Pacific Alaska's route. Lockheed Aircraft Corporation, Burbank, California

Eastern Air Lines had its own fire department. Yes, really. This photo comes from a woman whose grandfather is second from the right on the tail end running platform. From the DC-3 registration, it's clear that the photo was taken sometime in 1949 or 1950. (Photo Courtesy Laura Cortine Collection)

Pearce

Hardware on the Wing

Wright R-1820 Cyclone

As mentioned previously, the first high-power radial built at Wright was the R-1, designed in 1920 and tested in 1921. The 9-cylinder R-1 had a 5.625-inch bore and a 6.5-inch stroke. The engine displaced 1,454 ci and produced 350 hp at 1,800 rpm. The R-1 weighed 834 pounds. With R-1 development orphaned to Curtiss, the Wright team began to work on a new engine, the P-1.

The P-1 design originated with Charles Lawrance, and it was the first engine to carry the Cyclone name. Development of the P-1 began in 1923 and was supported by the navy. The P-1 was a 9-cylinder engine with a 6.0-inch bore and a 6.5-inch stroke. The engine displaced 1,654 ci and produced 400 hp at 1,650 rpm. The 812-pound engine had been designed with the smallest diameter possible. To achieve this, the intake valve was positioned at the rear of the cylinder and the exhaust valve was at the front. Both valves were actuated by pushrods at the front of the engine. The complex valvetrain proved to be unsatisfactory, but the navy was receptive to design changes that would enhance the engine's per-

The Wright P-2 showed some promise as a successful aircraft engine, but it was underpowered. However, the engine laid the foundation for what became the most successful of Wright's aircraft engines. (Photo Courtesy Aircraft Engine Historical Society)

formance and cure the valvetrain issues. The engine redesign led to the P-2 Cyclone.

Development of the P-2 began in 1924. The engine had the same bore and stroke as the P-1, but the cylinders were updated with standard valves and design features that had been engineered by Sam Heron while he was working at the Army Power Plant Lab. The P-2 was the first Wright engine to incorporate a supercharger, even though its primary purpose was to mix the air and fuel and not provide boost. Development of the P-2 was delayed due to the departure of Frederick Rentschler, George Mead, Andy Willgoos, and other Wright personnel. Tested in 1926, the P-2 produced 435 hp at 1,800 rpm and 500 hp at 1,900 rpm. The engine weighed 851 pounds.

By the time it was tested, the P-2 had been outshined by the P&W Wasp and Hornet. A larger engine was needed, and the P-2 was completely redesigned by Edward T. Jones and Sam Heron. Both men had joined Wright in 1926 to fill positions that had been vacated by engineers moving to P&W. The redesigned engine had the same bore as the P-1 and P-2, but the stroke had been length-ened by 0.375 inch and was now 6.875 inches. The changes led to a displacement of 1,749 ci, and the engine was known as the Wright R-1750 Cyclone. The R-1750 initially produced 525 hp at 1,900 rpm. The direct-drive engine had a diameter of 54 inches and weighed 760 pounds. R-1750 production began in 1927, and 568 engines were made.

On 9 August 1929, Wright merged with the Curtiss Aeroplane and Motor Company to create the Curtiss-Wright Corporation. For a short time, the development and production of Wright and Curtiss engines continued unabated. The Great Depression was the catalyst for a number of financial difficulties that put Curtiss-Wright on a path toward bankruptcy. At the same time, P&W powered 90 percent of the nation's air transports. Wright was losing market share and needed a plan to quickly reassert itself. In 1931, the separate Curtiss and Wright engine departments were merged and based out of the plant in Patterson, New Jersey. Development and production of the Curtiss engines were dropped, and Wright focused on

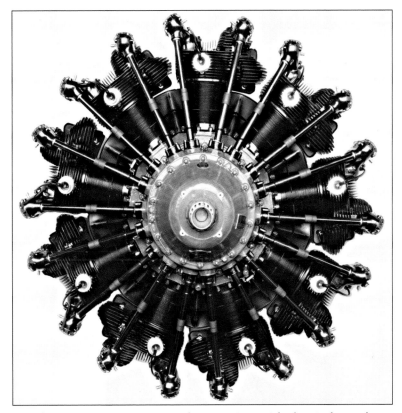

As an enlarged P-2, the R-1750 was another important developmental step toward the R-1820. Unlike in the P-2 and R-1820, exhaust from the R-1750 was expelled via a port on the front of the cylinder into a collector ring (not pictured) that surrounded the engine. (Photo Courtesy Aircraft Engine Historical Society)

The first R-1820 engine was the E-series, with the A through D designations applied to subtypes of the R-1750. Continual cylinder development is evidenced by the ever-finer and more numerous cooling fins. A finned, rear-facing exhaust port protruded from the cylinder. R-1820E production was around 721 engines. (Photo Courtesy Aircraft Engine Historical Society)

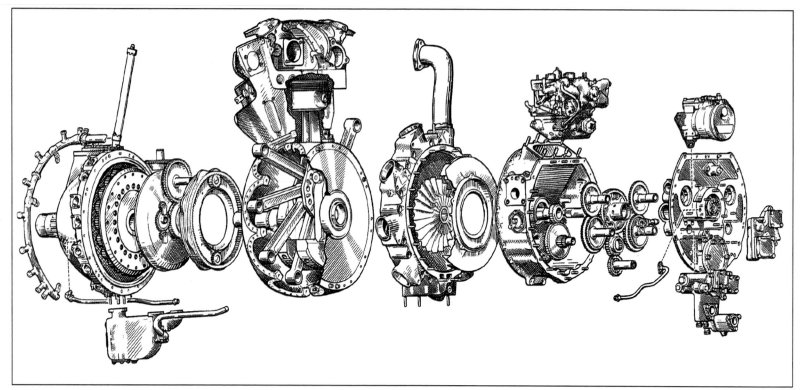

This exploded view of an R-1820 engine illustrates the complexity of a radial engine. On the left is the propeller reduction gear and cam ring. At center is the crankcase (with a single cylinder attached to the master rod) followed by the supercharger. On the right is the accessory section (with downdraft carburetor) and rear cover. (Photo Courtesy Illustrated Parts Breakdown Navy Model R-1820-80 Aircraft Engines (Curtiss Wright), *July 1961)*

its Whirlwind and Cyclone engines. The Curtiss division of Curtiss-Wright developed and produced aircraft while the Wright division developed and produced aircraft engines.

In 1930, the R-1750's bore was enlarged by 0.125 inch to 6.125 inches, but the stroke remained unchanged at 6.875 inches. The larger bore increased the 9-cylinder engine's displacement to 1,823 ci, and the Wright R-1820 Cyclone 9 was born. The first of the R-1820s was the E-series, with A- through D- being applied to various versions of the R-1750. The R-1820 E-series used a barrel-type, aluminum crankcase. The cylinders were made of cast-aluminum with integral rocker arm housings and machined cooling fins. The two valves per cylinder operated in a hemispherical combustion chamber, and the exhaust valve was sodium-cooled, a design pioneered by Heron. A one-piece master connecting rod, located in cylinder number-1, was used on a two-piece crankshaft.

Further development of the Cyclone 9 engine series brought about the R-1820 F in 1932. Around 3,673 engines of the type were produced. With its R-1860 Hornet B falling short, P&W never had a single-row radial engine that matched Wright's R-1820. (Photo Courtesy Aircraft Engine Historical Society)

This drawing of an R-1820 G-series cylinder illustrates its deep cooling fins. The steel barrel, with its own cooling fins, was screwed and shrunk into the aluminum head. On the right, the hollow exhaust valve was filled with sodium to draw heat from the valve's head, up its stem, and dissipate the heat into the cylinder's fins. (Photo Courtesy Lubrication Instructions for the Wright Whirlwind Aviation Engine, July 1927)

Combined cooling area of over 24 square feet (22,300 cm²)

The R-1820 G-series was produced throughout World War II. This is a partial cutaway of a G100-series engine. The master rod is in the upper cylinder; note the visible pinion of the planetary reduction gear. The scoop above the engine provided air to the downdraft carburetor. (Photo Courtesy Aircraft Engine Historical Society)

In the Wright engines, the crankshaft sections were joined by a clamping collar that slid onto an extension of the crankpin. A bolt pinched the collar tight to secure the connection. The engine was available with or without a 0.633 planetary gear reduction for the propeller. A General Electric supercharger was used; its purpose was to mix the air and fuel rather than to provide boost. Initially, the R-1820 E-series had a 5.1:1 compression ratio. On 80-octane fuel, the engine produced 595 hp at 1,900 rpm for takeoff and 575 hp at 1,950 rpm for normal operation. The engine had a 54.25-inch diameter and weighed 850 pounds for the direct-drive version and 920 pounds for the geared version.

Development of the R-1820 continued, resulting in the F-series of 1932. The cylinder design was improved, using a cast-aluminum cylinder head that was screwed and shrunk onto a steel cylinder barrel with machined cooling fins. The configuration increased the cooling fin area. The process of constructing a screwed and shrunk cylinder involved heating up the cylinder head and, in some cases, cooling the cylinder barrel. The male threads of the barrel were screwed into the female threads in the hot cylinder head, producing an interference fit. As the head cooled, it shrunk, locking the cylinder head to the cylinder barrel.

The F-series used a forged aluminum crankcase split on the cylinders' centerline. The speed of the supercharger's impeller was increased to provide additional boost and true supercharging of the incoming mixture. Special dynamic counterweights were developed to eliminate crankshaft vibrations encountered with increased engine power and adjustable-pitch propellers. The F-series was available in direct drive or with a 0.6875 or 0.625 propeller gear reduction. The engine had a compression ratio of 6.4:1. With 87-octane fuel, the R-1820 F-series produced up to 900 hp at 2,350 rpm for takeoff and 760 hp at 2,100 rpm for normal operation. The engine weighed around 1,000 pounds.

The power and reliability of the R-1820 F enabled Douglas to create its twin-engine DC-1 and DC-2 aircraft. The original request submitted to Douglas called for a trimotor airliner. By using more powerful engines, Douglas was able to meet the aircraft's design goals with just two engines, resulting in the DC-1 design. The use of just two engines reduced complexity, saved weight, extended range, and cut expenses. In addition, the twin-engine design enhanced cabin comfort by eliminating the noise, vibration, and fumes created by an engine mounted to the front of the fuselage.

Starting in 1935, the G-series further advanced the Cyclone engine with a slight compression ratio increase to 6.45:1. The G100-series followed in 1937 and began to introduce significant changes to the R-1820 engine. Cylinders were further refined, incorporating around 2,800 square inches of cooling fin area per cylinder. The aluminum pistons had cooling fins machined into their inner walls to take full advantage of internal oil cooling, and compression ratio was increased to 6.7:1.

The General Electric supercharger was replaced with a Wright-designed unit that was 25 percent more efficient, and a two-speed supercharger was available. The aluminum crankcase was discarded in favor of a specially designed, thin-wall, steel forging. The steel crankcase was approximately the same weight as the aluminum crankcase but much stronger and more rigid.

At the Douglas plant during World War II, rows of R-1820 engines were fitted to engine mounts and readied for installation on Douglas SBD Dauntless dive bombers. Around 100,000 examples of the engine were built during the war, mostly to power the Boeing B-17 Flying Fortress. (Photo Courtesy NARA 520741)

Crankcase thickness varied by location; the forged steel crankcase was around 0.25 inch thick, and the aluminum forging was around 0.75 inch thick. Magnesium castings replaced the aluminum castings used for the propeller gear reduction housing, two-piece supercharger housing, and accessory drive. The G-100 series was available in direct drive or with a propeller gear reduction of 0.6875 or 0.666. Using 91-octane fuel, the R-1820 G100-series engines produced 1,100 hp at 2,300 rpm for takeoff and had a normal rating of 900 hp at 2,300 rpm.

Introduced in 1939, the G200-series had 3,500 square inches of cooling fin area, which included around 1,100 square inches of fins that were 0.025 inch thick and machined into the steel cylinder barrel. Direct-drive versions of the engine were available, but most used either a 0.666 or 0.5625 propeller gear reduction. The engine had a takeoff rating of 1,200 hp at 2,500 rpm and a normal rating of 1,000 hp at 2,300 rpm using 100-octane fuel. The R-1820 G had a diameter of approximately 55 inches and weighed approximately 1,310 pounds.

The last production version of the R-1820 was the H-series (C9H, for Cyclone 9-cylinder H-series), and it entered production in 1942.

The engine used forged aluminum cylinder heads and had a 6.8:1 compression ratio. The cooling fins machined into the cylinder barrel were replaced by a series of small grooves. Cooling fin pairs made of thin, high-strength aluminum sheets were sandwiched together and rolled into the small grooves. Called the "W" cooling fin, the process originated with a single aluminum sheet that made three tight 180-degree bends.

The bent end formed a W and was rolled into the groove in the cylinder barrel. These fins were prone to cracking, and an improved W fin was created using the sandwiched cooling fin pairs. The fin no longer formed a W, but the name was retained.

Like the G200, the H-series were available in direct drive or with a propeller gear reduction of 0.666 or 0.5625. With 115/145 PN fuel, the engine produced up to 1,525 hp at 2,800 rpm for takeoff and 1,275 hp at 2,500 rpm for normal operation. The last R-1820 engines had a diameter of 55.75 inches and weighed around 1,425 pounds. A few engine models were outfitted with anti-detonation injection to increase their takeoff power by up to 150 hp.

The R-1820 was used in fighters, bombers, transports, helicopters, and tanks. During World War II, licensed R-1820 production

Wright R-1820 Engines in Airliners				
Manufacturer	A/C Model	Engine Model	HP t/o	Number
Boeing	307	GR-1820-G102	1,100	4
	307, B	GR-1820-G105	1,100	4
Curtiss	T-32 Condor II	GR-1820-F3	700	2
Douglas	DC-1, DC-2	GR1820-F2	660	2
		GR1820-F3	710	2
	DC-2	GR1820-F52	875	2
		GR1820-F53	770	2
		GR1820-F62	900	2
	DC-2, DC-3	GR-1820G2	800	2
	DC-3	GR-1820G5	1,000	2
		R-1820-49	975	2
		GR-1820-G102	1,100	2
		GR-1820-G103	1,000	2
		GR-1820-G105	1,100	2
		GR-1820G202A	1,200	2
Lockheed	14-WF62	GR1820-F62	900	2
	14-WG3B	GR-1820G3	875	2
	18-40	GR-1820-104A	1,200	2
	18-50	GR-1820202A	1,200	2

Major Applications of the Wright R-1820 Cyclone Engine	
Boeing	B-17 Flying Fortress bomber
	Model 307 airliner
Brewster	F2A Buffalo fighter
Curtiss	CW-21 fighter
	P-36 Hawk fighter (various versions)
	SBC Helldiver dive bomber
	SC Seahawk scout seaplane
	T-34 Condor II
Douglas	A-33 attacker
	B-18 Bolo bomber
	SBD Dauntless dive bomber
	DC-2 airliner
	DC-3 airliner (early versions)
	DC-5 transport
FMA	AeMB.2 light bomber
Grumman	C-1 Trader transport
	E-1 Tracer carrier warning aircraft
	FF / SF fighter
	F3F fighter
	F4F Wildcat fighter (various versions)
	HU-16 Albatross amphibian
	J2F Duck utility amphibian
	S-2 Tracker anti-submarine warfare aircraft
Lockheed	Model 14 Super Electra airliner
	Model 18 Loadstar transport
	A-28 Hudson recon bomber
Martin	B-10 bomber
North American	O-47 observation aircraft
	P-64 fighter
	T-28 Trojan trainer
Northrop	YC-125 Raider transport
Piasecki	H-21 Shawnee cargo helicopter
Ryan	FR-1 Fireball mixed-power fighter
Sikorsky	H-34 Seabat/Seahorse/Choctaw helicopter

was undertaken by Studebaker, and that company produced 63,789 engines. A diesel version of the engine, known as the RD-1820, was built by Caterpillar. The RD-1820 produced 450 hp at 2,000 rpm, and the complete engine package weighed 3,900 pounds, including the cooling fan and its drive clutches. Only 120 were built.

Wright continued to develop the R-1820 after the war, designing the turbo-compound TC9J2 in 1951. This engine had two power-recovery turbines, each feeding about 110 hp back to the engine. With 115/145 PN fuel, the Turbo Cyclone 9 produced 1,750 hp at 2,800 rpm for takeoff, a significant power increase of nearly 295 percent from the R-1820 E's original 595 hp at 1,900 rpm. Power for normal operation was 1,300 hp at 2,500 rpm. The TC9J2 had a diameter of 55.9 inches and weighed 1,750 pounds. The R-1820 Turbo Compound never entered production.

In the late 1950s, Lycoming and P&W Canada produced the R-1820 under license. Lycoming built 7,015 engines and P&W Canada built at least 287. The R-1820 engine stayed in production until the mid-1960s and was the last piston engine built by Wright Aeronautical. Approximately 120,000 R-1820 Cyclone engines were built.

The Wright R-1820 Cyclone 9 engine was used in more types of production aircraft than perhaps any other piston engine made in the United States. Each of the R-1820's cylinders displaced 202.6 ci, and no other radial engine achieved success with larger cylinders. By all accounts, R-1820 engines gave reliable service when established operational and maintenance practices were followed.

This old workhorse DC-3 freighter, dressed in modern cargo carrier livery, is flying from Northern to Southern California in 1972, perhaps by our author. (Photo Courtesy Robert Campbell Aerial Photography)

by John K. Lewis

Although two military assignments flying C-47s exposed me to P&W R-1830 Twin Wasp engines, it wasn't until after retirement from the air force, while flying freight in DC-3Cs, that I had an opportunity to experience both R-1830 Twin Wasp and Curtiss-Wright R-1820 Cyclone 9-powered aircraft. TransWest Air Express flew scheduled weekday routes from the home base in Oakland to Los Angeles, Phoenix, and Salt Lake City using three aircraft each night.

As one DC-3 left Oakland for Los Angeles and Phoenix, a second departed Phoenix for Los Angeles and Salt Lake City, and the third took off from Salt Lake bound for Los Angeles and Oakland. Later in the year, Portland, Oregon, was added to the schedule, and eventually a DC-3 was positioned at LAX when freight volume grew to a point where two aircraft were needed to handle loads between those cities. Each night the southbound DC-3 from Oakland and northbound bird from Los Angeles passed each other near Paso Robles.

Flying freight at night in the DC-3 became a delicate balance of moving a maximum amount of freight on each leg while pampering the engines to extract as many hours as possible before overhaul. This took some planning because the DC-3 was a "wing" airplane in contrast to something like a Lockheed F-104 Starfighter, which was *in every way* an "engine" airplane. Descents into LAX were the most difficult. Pilots started requesting lower altitudes early but invariably were delayed by traffic and found ourselves struggling to get down to approach altitudes.

Round engines last longer if you keep a positive load on the thrust bearings, so instead of just retarding throttles to idle, we had the propeller controls all the way back to keep manifold pressures above the existing propeller RPM. This was the only way we could ensure positive loading on the bearings while obtaining maximum descent rates. Los Angeles Center, SoCal Approach Control, and even the tower at LAX were all constantly after us to increase the rate of descent. In addition, the Federal Aviation Administration (FAA) mandated carrying a maximum of 5,000 pounds of freight as payload on any one leg. DC-3s could carry more, a whole lot more, but couldn't maintain altitude if an engine failed at higher gross weights.

My logbook doesn't reflect the actual dash numbers of the R-1820 and R-1830 engines, so I can't identify which of the 50 different models of Wrights and 33 different models of Pratts were in the fleet. Still, some of the Pratt-powered aircraft seemed more robust than others, and the one Wright R-1820-powered aircraft must have had later-model Cyclones because it performed better than all the others. There also didn't seem to be any difficulty keeping the twin-row Pratt cylinders cool versus the single-row Wrights. Cowl flaps were left in "trail" during climb, and were closed after reaching cruising altitude once the engines had cooled down.

The Pratt R-1830-94 is shown on a C-47. (Photo Courtesy Wikipedia Commons)

Carburetors had three settings: Auto Rich, Auto Lean, and Idle Cutoff. Takeoffs and climbs were performed in Auto Rich while cruise flight used Auto Lean. Incidentally, DC-3 carbs didn't offer a method of manually leaning mixtures. The Wright R-1820s consumed more fuel per hour than the P&Ws, so a 10-percent increase was added to the P&W fuel-burn charts when pilots calculated fuel.

The maximum number of hours pilots could fly each month was 100 along with a maximum of 1,000 hours per calendar year. To achieve the maximum number of hours out of the pilots as well as the DC-3's engines, Trans West had the pilots come close to the 1,000-hour maximum. Along with that, the pilot experience was, for the most part, flying all night or night weather time without autopilots.

Although I had developed a fondness for P&W engines before joining TransWest, it was obvious that the Wright Cyclone engines were more powerful. Granted, they didn't run as smoothly as the 14-cylinder Pratts, and with only 9 cylinders each, the Wrights couldn't compete for smoothness. Still, they had fewer moving parts to fail, so it was a trade-off, and the freight never complained about the ride.

The Wright Cyclone S1C3-G engine on a TWA DC-3 is shown. (Photo Courtesy Museum of Flying Collection)

Chapter 3

PRESENTING THE SHIP AS AN AIRPLANE

Perhaps no other time period in American cultural history is steeped in such absolute romance as the 1930s. So much of this allure had to do with the aspects of "seeing the world," and we as a people were never more aware of the outside world than during this period following our excursion overseas to discover Europe during the Great War.

Radio was connecting us as never before, and motion pictures of exotic overseas locales were available in the form of movie news reels, talking pictures having become a vital part of everyday life in America. This of course spurred the rich and adventurous to want to see this new world for themselves and be swept-up in the kaleidoscope of sights, sounds and dangers of exotic far-off lands and cultures.

Getting There by Air

The 1930s

Flying high over San Francisco Bay is the fated Boeing 314 Pacific Clipper from "round the world" fame. Note the pre-World War II identification of a US aircraft via the American flag. (Photo Courtesy Boeing Co. via Jon Proctor with permission)

With the natural human tendency to explore, the newfound need to span the globe was facilitated by a new transportation system: the airplane. So much faster than a plodding ship, you could, by 1935, travel from San Francisco to Hong Kong in about five days, instead of by sea in three weeks. Once again, however, it took a visionary man to make such a transportation system viable and practical. The world was given Juan Trippe and his Pan American Airways (PAA) as the answer to the challenge.

Trippe was a truly driven and ruthless businessman, as larger-than-life leaders often are. Starting in 1929, he began amassing airline companies and their airmail contracts with countries other than the United States as well as those from our own Post Office Department. His first routing system grew from Key West and Miami to Havana, Cuba, to eventually encircle the entire Caribbean, including the coastal regions of Mexico, Central American nations, and the northern grouping of sovereigns on the South American continent.

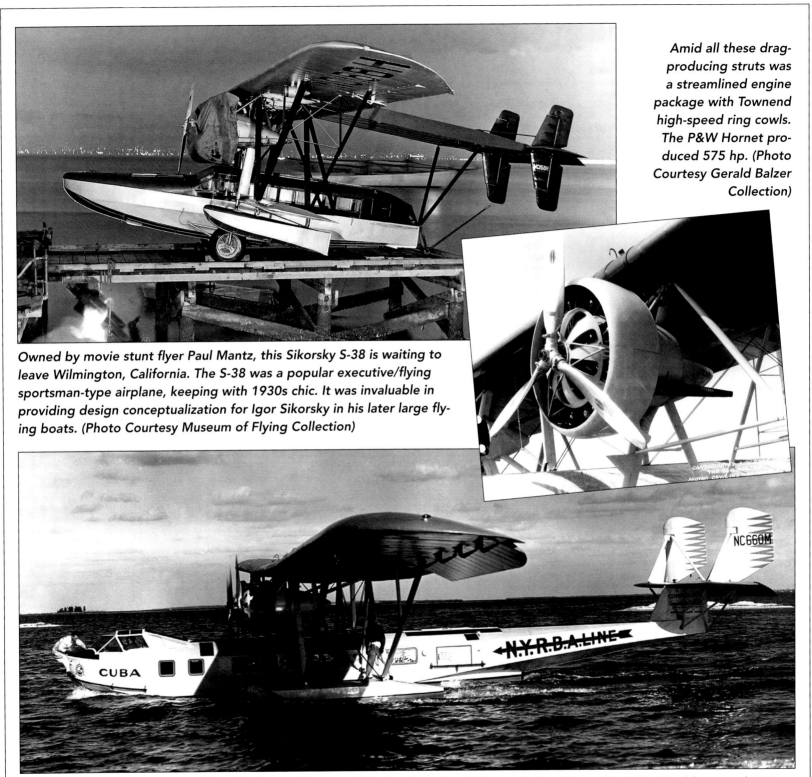

Amid all these drag-producing struts was a streamlined engine package with Townend high-speed ring cowls. The P&W Hornet produced 575 hp. (Photo Courtesy Gerald Balzer Collection)

Owned by movie stunt flyer Paul Mantz, this Sikorsky S-38 is waiting to leave Wilmington, California. The S-38 was a popular executive/flying sportsman-type airplane, keeping with 1930s chic. It was invaluable in providing design conceptualization for Igor Sikorsky in his later large flying boats. (Photo Courtesy Museum of Flying Collection)

The seminal flying boat service launched Juan Trippe and his Pan American Airways System throughout the Caribbean and Latin America. Consolidated Commodores, originally designed for navy patrol flying, were perfect for Ralph O'Neil's NYRBA. After PAA acquired the line, the Commodore routinely flew more than 600 miles (the longest nonstop route in the world). (Photo Courtesy Gerald Balzer Collection)

Following this bout of building, Trippe set his sights on the successful NYRBA airline (New York, Rio, and Buenos Aires Line), which flew all along the eastern coastlines of the two continents, linking Miami with Argentina and Chile. After a gruesome set of maneuvers, Trippe emerged as the victor and now owned a successful and well-run airline company. PAA was now in high form and ready to take on the world.

Charles Lindbergh was a busy consultant in the late 1920s working, as we have seen, at TWA as well as PAA. His route-proving flights coincided with Trippe's judgment that amphibious aircraft were best suited to the rout system then emerging within PAA. NYRBA was also using the flying boat as its conveyance, choosing the Consolidated Commodore (Consolidated's owner, Rueben Fleet was co-owner of NYRBA). This is the vehicle that grew the romance of the decade: an ocean-going airplane that was configured much like an expensive yacht. PAA was flying the Sikorsky S-38, smaller than the Commodore but a true amphibian. This began a long relationship with the Connecticut airplane builder and PAA and proved to be pivotal to the progression of ocean-spanning routes by Trippe and company.

The airline was constantly expanding, including Alaska, China, and Latin America, including Mexico. But the jewels in the crown and those experiences of highest profile for Americans to note were the eventual routings from the United States to the Far East and United States to Europe, both segments flown by pure flying boats. The Sikorskys were the intermediate airframes applied but were limited in what they could accomplish. These early airplanes were

A Sikorsky S-40, whose lineage is directly attributable to the S-38. Now more than just a "flying yacht," the larger S-40 was truly capable of hauling paying passengers. (Photo Courtesy Gerald Balzer Collection)

The way to stabilize a flying boat in the water always came down to pontoons or sponsons. Consolidated chose kind of a hybridized version, designing a "spon-toon" if you will, which looked like a bass boat attached to the airplane. (Photo Courtesy Gerald Balzer Collection)

Flying to Rio or Havana, life aboard a PAA Clipper was high-end living at its best during the Depression. Traveling to warm tropical climates required wearing light-colored, lightweight suits for men and summer dresses for women.

The Sikorsky flying boat matures. Here is the S-42 "on the step" leaving Dinner Key in Miami, PAA's main base. Obvious streamlining is evident in contrast to the earlier S-40, including extensive flush riveting. Large wing flaps and prop brakes were also new. (Photo Courtesy Gerald Balzer Collection)

This interior of the S-42 highlights the extreme crossover nature of the airplane turned boat. All the usual nautical touches are present, creating a feeling of a world-class yacht (of immense size, of course). The woodwork is impressive and substantial with "happy" fabrics on all the seating surfaces. (Photo Courtesy Gerald Balzer Collection)

the models S-40 and S-42 and really introduced the flying public to an "ocean liner of the skies" mental construct and overall service ambience.

The S-40 flying boat could carry 38 passengers and travel 900 miles. The first aircraft went into service in November 1931, and were named Clippers, capitalizing on the oceangoing merchant ships of yore. Thus began the most significant trademarking of equipment perhaps in American history.

The first airplane so named was appropriately *American Clipper*. Only three were built, but they were the literal building blocks for

PAA transoceanic service. Remember that the land planes of the time didn't carry as many fares as this flying boat (but not by a large number). These were significant airplanes. Onboard service was by steward in a white tuxedo waistcoat and included china and silverware, just as you found at home or at the Waldorf Astoria.

Next came the much-improved S-42. Here was the true refinement of the S-38/S-40 archetypes. This airplane was coincident with the Douglas DC-2 and had as many innovative and up-to-date aeronautical features. It could carry 32 passengers, was heavier, and flew farther than the landplane. It was used throughout the Caribbean,

Beginning in April 1935, PAA flew an S-42 on route-proving flights all the way west to Manila. These early flights paved the way for routine service across the Pacific by the soon-to-be-delivered Martin M-130 Clippers. Upon completion of these flights, the US Postal Service awarded PAA Airmail Route 14 spanning the Pacific Ocean, paying $2 per mile to deliver mail. Here, the Sikorsky is approaching Honolulu with Diamond Head in the background.

Latin America, and South America. Awaiting the larger airplanes designed for the transoceanic gambits, Trippe modified one S-42B to fly only PAA personnel on route-proving flights across the Pacific and Atlantic oceans. This airplane provided the information for the stepping-stone route system that made Pan American Airways a household name imbued with mystique just a year and a half later.

Conquering the Pacific

Trippe asked both Sikorsky and Martin to propose a long-range oceangoing Clipper flying boat that could easily cross the great wet expanses. The Martin design received the nod and was known as the M-130. PAA chose to name the set of Martins as *Hawaii Clipper, Philippine Clipper,* and most iconic and famous of all, *China Clipper.* Each airplane cost the company $417,000, could accommodate up to 42 passengers, and had a range of 3,200 miles.

Romance aside, the PAA system was exciting and adventurous but always operationally sound and conservative. To accomplish crossing the pacific to China with available technology in 1935 required deft planning and technical acumen. PAA's answer to the question of how to cross all that water relied on nature's gift to the airline: the pacific islands.

Using a series of islands that stretched from California to the Far East, the long route west could be nibbled at one segment per day. Each one of the islands served as both airplane and passenger "refueling" stops. Trippe spent a small fortune designing and building infrastructure on each of the islands, creating "stations" for PAA personnel and passengers. Supplies were brought by ship to each island outpost.

The airplane for the Pacific was Martin's sleek M-130. Three Martins were ordered, being christened by PAA as Hawaii Clipper, Philippine Clipper, and most famously of all, China Clipper. The first passenger flight was on 22 November 1935 from Alameda, California, captained by Ed Musick. The navigator on that flight was none other than Fred Noonan, later to leave PAA and become navigator for Amelia Earhart on her ill-fated round-the-world flight.

PAA workers built small cities to accommodate flight ops and guests. Hotels with accompanying swimming pools were built, radio and weather observation and reporting buildings were erected,

PAN AMERICAN AIRWAYS Martin M-130 *"China Clipper"*

Pre-Construction Artist's Impression - 1933

Specifications & Performance

Manufacturer: Glenn L. Martin Company – Baltimore, Maryland
First Flight: December 30, 1934
Service: Pan American Airways, Inc.
Number Built: 3
    Hawaii Clipper (NC 14714)
    Philippine Clipper (NC 14715)
    China Clipper (NC14716)
Cost: $417,000
Crew: 6-9 (Captain, First Officer, Junior Flight Officer, Engineering Officer,
    Assistant Engineering Officer, Radio Operator, Navigation Officer &
    Cabin Stewards)
Passengers: 46 'Day' Flights (between islands)
    8 'Overnight' Flights (San Francisco - Hawaii)

Wingspan: 130 ft
Length: 90 ft 7 in
Height: 24 ft 7 in
Empty Weight: 29,013 lbs
Maximum Gross Takeoff Weight: 52,252 lbs
Endurance: 20+ hours
Range: 3,200 nm
Service Ceiling: 10,000 ft
Fuel Capacity: 4,080 gals
Maximum Speed: 180 mph
Cruise Speed: 130 mph

Engines: 4 × Pratt & Whitney R1830-S2A5G Twin Wasp 14-cylinder air-cooled
    radial engines rated at 830 hp (later 950 hp with hydromatic propellers)
Propellers: 3-Bladed Hamilton Standard Hydromatic
Aircraft Losses: All 3 M-130s were lost in accidents.
    Hawaii Clipper disappeared without a trace over the Western Pacific Ocean enroute
    from Guam to Manila on July 28, 1938. 15 fatalities.
    Philippine Clipper struck a mountain near Ukiah, California enroute from
    Hawaii to San Francisco on January 21, 1943. 19 fatalities.
    China Clipper struck the water and broke up on approach to Port of Spain-Léopoldville
    enroute from San Juan on January 8, 1945. 23 fatalities.

Legend (from GLENN L. MARTIN COMPANY Press Release - 1933)

(1) Captain & First Officer
(2) Radio Operator
(3) Engineering Officer
(4) Mooring Compartment
(5) Passenger's Luggage, Mail & Express Hold
(6) Forward Hatch
(7) Galley
(8) Steward
(9) Navigation Officer & Plotting Table

(10) Crew Rest Area & Bunks
(11) Captain's Desk
(12) Passenger's Lounge (15 passengers)
(13) Passenger Compartments (2 positions)
    For Shorter Daylight Flights Between Islands - 20 lounge chairs
    For Longer Overnight Flights - 12 sleeping berths
(14) Lavatory & Dressing Rooms
(15) Additional Cargo Hold / Emergency Equipment
(16) Crew Bunks

(17) Water Tank
(18) Radio DF Mast
(19) Sponson (sea wings - port & starboard)
(20) Fuel Tanks (2 positions in keel)
(21) Fuel Pump (in keel)
(22) Pratt & Whitney R1830-S2A5G Twin Wasp 14-Cylinder
    Air-Cooled Radial Engines (4 positions)
(23) 3-Bladed Hamilton Standard Hydromatic (4 positions)
(24) Pitot Static Mast (for airspeed indication)

This is a fine color study of the interior arrangement and amenities of the Martin M-130.

Aloha and welcome to Hawaii! Note that just like the Sikorsky boats, the Martin required entering the fuselage through a top-located hatch. Note the stairway contraption to get passengers up and over the airplane.

The docking and birthing arrangement at Pearl City, Honolulu, Hawaii. PAA's operations were located at the far north end of Pearl Harbor.

When you are first, all the infrastructure is up to you, and so it was with PAA. The company built identical hotels at all island stopping points, this one located on Midway Island (note the indigenous albatrosses, aka gooney birds). Every stick of lumber and any support equipment had to be delivered by ship because the islands were nearly deserted and without resources. Such was pioneering aviation in the 1930s.

water collecting and desalinization centers were created, and piers and hardscape were placed as necessary. Even coral heads had to be detonated in the Wake Island lagoon so as to make a safe landing area for the flying boats. Talk about building from the ground up!

Flight operations were also cutting edge, and because no structure was in place on most of the routing, PAA had to do it/bring

ROUTE OF THE CHINA CLIPPER

it/build it/design it itself. And although the Martin boats were designed specifically for the Pacific, they still operated based on the whims of the weather, true of all flying operations, even today.

Obviously, range was the largest and most abiding consideration for the Pacific portion of PAA's system, especially the longest route segment: California to Hawaii. So critical was flying west in the winter months that it was not uncommon to be able to transport only a few paying passengers when the winds out of the west/northwest were up. In one famous incident, a single woman passenger was allowed to board the flight, range being so critical that all the additional weight was to be consumed by fuel load. The airplane departed the Bay Area, could only fly at 150 feet above the waves to stay below the worst of the storm winds, and eventually had to return to Alameda anyway because it was clear that they simply could not make it all the way to Hawaii. This was an example of cutting-edge technology; pioneering is hard and fraught with extreme dangers.

For the most part, however, flying by Clipper was fairly uneventful, again a tribute to PAA's tight operating procedures. Captains, for instance, had to be rated as both pilots and navigators. They also had training in radio operations and nautical knowledge. Plenty of staff were always available at each base to ensure trouble-free service.

All aspects of traversing the open ocean and alighting on desert islands were very civilized. Passengers arrived at the dock at Wake or Midway, were shuttled the few yards to the hotel, and offered cocktails in the open-air lounge, warm ocean breezes filtering through the atmosphere. Time for a dip in the pool was scheduled prior to dinner service then a stroll around the island or cards in the lounge before retiring and setting off the next day for yet another mysterious island location. This was island living at its gracious best and is what made PAA the preeminent international airline in the world.

Inflight service was as during the preceding years on the Sikorskys, but it was a touch more grand and refined now. The M-130s allowed for more room to stretch and relax or visit. Best

Look at this menu from PAA's flight 263 on its first leg west. Where did they have the room for all this gourmet food onboard?

PAA

PHILIPPINE CLIPPER
Capt. McGlohn   *R. H. M. Glohn*
MENU

FLIGHT No. 263   Jan. 21, 1939
Flight Steward Merrill

DINNER

Chilled Celery Hearts   Mixed Olives
California Fruit Cocktail

Clear Consomme
Steamed Spring Chicken, Fricassee Sauce
Parslied New Potatoes   Fresh Garden Peas

COLD BUFFET:

Roast Prime Ribs of Beef   Liverwurst Sausage
Baked Virginia Ham

Avocado and Grapefruit Salad,
French Dressing

Neapolitan Ice Cream   Pound Cake

American, Bondo, and Phil. Cream Cheese,
Ritz Butter Wafers

Coffee   Tea   Milk

Assorted Fresh Fruit   Brandied Dates
Salted Nuts

FROM Alameda   TO Honolulu

of all, you could be at the destination of your choosing, whether Honolulu, Manila, or Macao, in a matter of hours or days, not weeks. The world was beginning to shrink, noticeably.

Go East, Young Airline

As good as the Martin Clippers were, Juan Trippe had one more stretch in flying boat design growth to accomplish. His engineers settled on Boeing to follow their requirements for the world's largest and longest-range flying boat to date. Each airplane weighed 82,000 pounds fully loaded and could fly 3,500 miles unrefueled. Boeing called its airplane the Model 314. Everyone else called it huge.

The 314 flew the Pacific first, as far south as Auckland, New Zealand. By this time, the airline had opened a new sprawling seaplane base on Treasure Island across the bay from Alameda and in the shadow of the Bay Bridge connecting San Francisco to Fremont. This was PAA's West Coast headquarters and gave credence to its grandiose self-image and public image.

California Clipper at rest at Pearl City. This 314 had the unenviable but dramatic distinction of flying completely around the world from east to west the week that America was plunged into World War II following the attack on Pearl Harbor. With the crew improvising and virtually holding the ship together with bailing wire and tape, it made it home to LaGuardia in New York after having flown more than 31,000 miles. Note the American flags painted conspicuously on the airplane prior to our direct entry into the war. At this time, the United States was neutral in the conflict. (Photo Courtesy Jon Proctor)

Not long after, on 20 May 1939, transatlantic service began from Port Washington, New York. The inauguration of this route gave PAA the distinction of being the first airline in the world to run scheduled airline service across the Atlantic Ocean. It was 12 years after Lindbergh. Two routes were used, a northern segment that dotted Newfoundland, Canada, and a southern segment via the Azores in the mid-Atlantic. If taking the trail north, the airplane would make first landfall at Foynes, Ireland. In the typical cold and damp of the North Atlantic weather, passengers were whisked into the arrival building from the dock and given a hot libation to warm them sufficiently before returning to the air and proceeding to Southampton, England. The original drink concocted by the local inn keeper consisted of hot coffee with a shot of good Irish whisky. This became known worldwide as, you guessed it, Irish Coffee.

A pause now to acknowledge with all possible emphasis that what was occurring at PAA in 1939 was nothing short of stupendous. Thanks to the precision of operating procedures, the technology of aerial navigation pioneered by PAA, long-range aerial communications pioneered by PAA, and the industrial capability and innovation of the modern engine manufacturers and airframers challenged by Trippe, Pan American Airways was the envy of the world's airlines. The company had made routine what was considered just a few years prior to be unassailable by man or machine.

Bear in mind also that these large strides in flyability of the world's geography were made possible by the fact that only ocean-going aircraft could lift the corresponding loads required to make the distance. They had an entire ocean to use as a runway, so performance from landing strips was not the consideration that it is for land planes. This was all about to change, but in 1939, on a day-to-day basis, flying boats were kings of the hill, and airlines the world over, along with their aeronautical design companies, were thinking of larger and larger ocean-based airliners.

The outbreak of World War II on 3 September 1939 (just five months after transatlantic service began), put an end to the northern route segment for PAA. The central and Brazilian routes remained. The airplanes still flew throughout the conflict, setting records as well. Eventually, as America entered the war as a combatant, PAA airplanes and personnel were appropriated by the navy and the aircraft camouflaged for flying service.

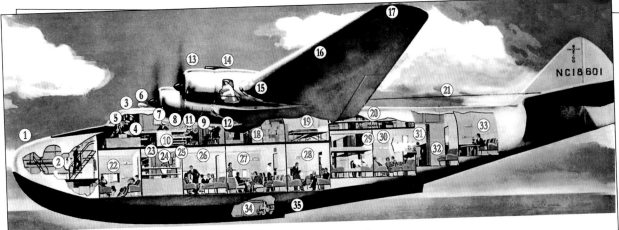
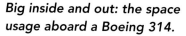

BOEING Model 314 *Clipper*

Pre-Construction Artist's Impression - 1937

Specifications

First Flight: June 7, 1938
Number Built: 12
Retired: 1951
Service: Pan American Airways, British Overseas Airways & U.S. Navy
Crew: 6-10
Wingspan: 152 ft
Length: 106 ft
Height: 28 ft
Gross Weight: 82,500 lbs

Empty Weight: 50,268 lbs
Engines: (4) Wright GR-2600-A2 "Double Cyclone" 14-cylinder
 Air-cooled radial engines rated @ 1600 horsepower/each
Maximum Speed: 193 mph
Cruise Speed: 183 mph
Cruise Ceiling: 13,400 ft
Maximum Ceiling: 19,600 ft
Climb: 565 feet/min
Range: 3,500 nm
Payload: 40-74 (36 Night) passengers

Legend *(from BOEING Press Release)*

(1) Anchor Hatch
(2) Seaman's Compartment
(3) Bridge
(4) First Pilot
(5) Second Pilot
(6) Radio Direction Finder Loop
(7) Navigation Compartment
(8) Radio Officer's Post
(9) Chart Room - Navigator's Post
(10) Bombs, Flares & Navigational Instruments
(11) Engineering Officer & Mechanical Engine & Aircraft Controls
(12) Captain's Office

(13) 1,600 horsepower engines with automatic adjusting propellers
(14) Mechanic's wing station
(15) Controllable landing lights
(16) Wing spread 152 feet
(17) Navigation lights
(18) Main cargo hold which extends into wing
(19) Crew's sleeping quarters
(20) Luggage holds
(21) Overall length of ship - 106 feet
(22) First passenger compartment, accommodations for 10
(23) Spiral staircase to bridge
(24) Men's retiring room

(25) Galley - Two stewards working simultaneously preparing food for 74 passengers
(26) Second passenger compartment, accommodations for 10
(27) Dining lounge, accommodations for 15 passengers
(28) Third passenger compartment, accommodations for 10
(29) Fourth passenger compartment, accommodations for 10
(30) Fifth passenger compartment, accommodations for 10
(31) Ladies dressing room
(32) Sixth compartment
(33) Private Cabin Suite
(34) Fuel pumps for transferring fuel from sea-wings to wing tanks
(35) Auxiliary hold

Big inside and out: the space usage aboard a Boeing 314.

After the end of hostilities, the 314 flew until 1946 when the New World Order of commercial aviation eclipsed the finest flying boat known to that moment. The days of ocean liner stateliness were receding, and although passengers began the postwar period looking forward to giant double-decked land planes circling the globe nonstop and bars and lounges ensconced within their capacious fuselages, the reality of Coach Fares and crowds was looming. Getting there by air was becoming an "everyman's" predilection.

Alameda was a temporary seaplane base for PAA only until it could construct a real airline operation across the bay on Treasure Island. As big as the Boeing is, it looks small compared to the architecture behind it. This was a premier bit of infrastructure in the world of intercontinental air transportation in the late 1930s. The Pacific Exposition was also held on Treasure Island, and many of the edifices behind PAA's facilities are part of that fair. During the war, the US Navy operated from this base, having appropriated PAA's flying boats. (Photo Courtesy Collection of SFO Museum, gift of Carla B. Bos in memory of Abraham Bos)

The 314 absolutely dwarfs visitors surrounding it on Treasure Island during an open house event. This is California Clipper, just prior to her adventure.

This is a rare color photo of the dining room portion of the Boeing 314. Refer to the cutaway drawing to see its location near the center of the fuselage and compare this scene to the card players aboard the Sikorsky S-40 earlier in this chapter. (Photo Courtesy Boeing Co. via Jon Proctor with permission)

Built expressly for Pan American by Pan American, the Marine Air Terminal at the new LaGuardia airport in New York was a tour de force in its interior design and embellishments. Huge murals paid for by the Works Project Administratrion (WPA) were painted all along the upper portions of the walls, soaring to 25 feet. As with all PAA terminals, the omnipresent giant globe of the earth was smack in the middle of the rotunda. The building is used to this day for Delta's East Coast shuttle between Washington, DC, New York, and Boston.

Pan American and the world it made.

This is a PAA flying boat captain. Flight officers were knowledgeable in flying, navigating, meteorology, radio telephony, and engineering. These were men of stature doing challenging work to push the boundaries of time and space around the globe.

A 314 is on the step departing Port Washington, New York. This was the base for Atlantic Operations until LaGuardia could be completed in 1939. (Photo Courtesy Museum of Flying Collection)

Pearce

Hardware on the Wing

Pratt & Whitney R-1830 Twin Wasp

As the design team at P&W developed the 575-hp Hornet B (R-1860), even larger and more powerful engines were contemplated. The 6.25-inch bore of the Hornet B caused some issues, and the development team preferred a smaller bore. With a smaller bore, the larger engine would need to have more cylinders. Positioning 11 cylinders around the crankshaft resulted in an engine with a very large diameter, and P&W was not interested in pursuing this option. A different engine configuration needed to be devised. P&W also needed to stay ahead of Wright and its R-1750 Cyclone engine. Although the R-1750 had a smaller displacement than the Hornet B, the Wright engine was proving to be quite remarkable.

In January 1929, Leonard (Luke) Hobbs was put in charge of designing an experimental engine to investigate a twin-row radial configuration. Hobbs had joined P&W in 1927 as a research engineer and had proven his worth. The idea for the experimental engine was simply to add another row of cylinders that was staggered behind the first row. Cooling air would pass between the cylinders of the first row and directly into the cylinders of the second row. This configuration would keep the engine's diameter small while essentially doubling its power. The concept of having two rows of cylinders was not new; it had been used in the pre–World War I days of aviation on semi-radials, small radials, and even rotary engines. Also, the two-row 14-cylinder Armstrong Siddeley Jaguar had been developed in Britain in the mid-1920s. However,

Hobbs's experimental engine would be the first twin-row design for P&W.

The experimental engine designed by Hobbs was designated R-2270, and it had two rows of seven cylinders. Each cylinder had a 5.75-inch bore and a 6.25-inch stroke. The 14-cylinder engine's total displacement was 2,272 ci. The cylinders were the same as those used on the Wasp engine, and other engine components came from both the Wasp and Hornet. The pushrods for the rear cylinders extended between the front cylinders. This configuration enabled a single cam ring at the front of the engine to actuate the pushrods for all 14 cylinders. The engine's design reverted to a one-piece crankshaft and split master connecting rods. On the first run in May 1930, the engine produced 870 hp and showed P&W the future of radial engines. Only one R-2270 engine was built.

The design of a new twin-row production engine was started in December 1930. Hobbs and Andy Willgoos were quick to apply everything learned from the R-2270 to this new engine. The engine was going to be more powerful than any other P&W engine built thus far, and a conservative approach was taken by using a moderate bore diameter and a moderate stroke length.

The new twin-row engine had a 5.5-inch bore and stroke. With its 14 cylinders, the engine had a total displacement of 1,829 ci and was 48 inches in diameter. The engine was called the Twin Wasp or R-1830. It had a similar layout to the R-2270, with one major exception: the R-1830 had two cam rings. One cam ring was positioned in the usual location, at the front of the engine. This ring actuated the valves of the first row of cylinders. A second cam ring was positioned at the rear of the engine and actuated the valves for the second row of cylinders. The cam rings had two tracks, each with four lobes. Each cam ring was driven in the same direction as the crankshaft and at 1/8 crankshaft speed.

The cylinders were made of a cast-aluminum head that was heat shrunk and screwed to a forged-steel barrel. The barrels had cooling fins that were machined into their outer sides. Because of the different valve drive, the heads of the front and rear cylinders were not interchangeable. As engine development continued, thinner and more numerous fins were cut into the cylinder head and barrel for better heat dissipation. Each cylinder had one intake and one sodium-cooled exhaust valve.

Cylinder numbering for twin-row radials followed the established practice of starting with the top cylinder and continuing clockwise when the engine was viewed from the rear. The rear row's top (vertical) cylinder was number-1, the number-2 cylinder was in the front row, and number-3 was in the rear row. This numbering convention placed all odd-numbered cylinders in the rear row and all even-numbered cylinders in the front row. Master rods for the R-1830 engine were located in cylinders number-5 and -12 except for the last versions of the R-1830 C-series, which had the master rods located in cylinders number-6 and -9.

A three-piece, forged aluminum crankcase was used that was composed of a front, intermediate, and rear section. Each section was split vertically through the cylinders. A nose case made of either cast aluminum or cast magnesium housed the propeller gear reduction. The use of magnesium helped reduce the engine's weight. Engines with propeller gear reduction ratios of .50 and .5625 used bevel gears, and a .667 reduction used planetary gears. Depending on the model, the rear accessory section was made of either cast aluminum or cast magnesium. The accessory section housed the single-stage single-speed supercharger, and the engine could be run with a turbosupercharger. Also, later engine versions incorporated a two-speed supercharger.

The R-1830 A-series was first run in April 1931 and was first flown in June 1931. The engine initially produced 750 hp at 2,300 rpm on 80-octane fuel with a 6.0:1 compression ratio. Using 87-octane fuel brought engine power up to 800 hp at 2,400 rpm. The R-1830 entered production in 1932 and had a takeoff rating of 950 hp at 2,550 rpm and a normal rating of 850 hp at 2,450 rpm. B-series engines soon followed, with a takeoff rating of 1,000 hp at 2,600 rpm and a normal rating of 900 hp at 2,450 rpm. A- and B-series engines weighed approximately 1,162 pounds.

By mid-1936, the first R-1830 C-series had been developed. Initially, these engines were not much more powerful than the B-series, but they were a stepping stone to greater performance. The early C-series engines were rated for 1,050 hp at 2,700 rpm for takeoff and 900 hp at 2,550 rpm for normal operation. A serious problem arose when the bearing between the master connecting rod and crankpin seized on the more powerful engines during a dive. In a dive, the engine was over sped to 3,000 rpm, and the added stress caused the bearings to fail. Bearing failures while in a steep dive was a condition that only occurred on military aircraft, but it was still a serious issue and something that needed to be resolved.

P&W initiated exhaustive testing of numerous alloys to find a bearing material that would not fail. A year later, the P&W team developed a process that yielded bearings that withstood 4,000 rpm. A thin (.002 inch) layer of a lead alloy (lead with 4 percent indium by weight) was electroplated over a thicker (.02 inch) layer of silver. The silver was bonded with copper to the hard steel bearing base. This silver-lead-indium bearing material used for the connecting rod was deemed so important that the method to create it was shared with other engine manufacturers in the United States and Britain as World War II approached.

By 1939, the C-series had been improved with a new supercharger and redesigned cylinders with a 6.7:1 compression ratio. These engines had a takeoff rating of 1,200 hp at 2,700 rpm and a normal rating of 1,050 hp at 2,550 rpm. The improved C-series weighed approximately 1,467 pounds, and some of the engines were fitted with a two-stage supercharger. Like Wright, P&W became disenchanted with General Electric's superchargers and began to develop its own. The two-stage supercharger used on the R-1830 was designed by P&W and used two impellers. The R-1830 was the first engine with two-stage mechanical supercharging applied to a production military aircraft when it was installed in the Grumman F4F-3 Wildcat in 1939.

The most powerful versions of the R-1830 C-series had a takeoff rating of 1,350 hp at 2,800 rpm and a normal rating of 1,100 hp at 2,600 rpm. The engine's weight had risen to 1,555 pounds. When first introduced, the R-1830 had a time between overhauls of 300 hours. As the engine was developed, this time was eventually stretched to 1,200 hours. Licensed production of the R-1830 during World War II included 74,198 engines built by Buick and 56,484 engines built by Chevrolet. R-1830 production continued until 1951, and a total of 173,618 engines were built. The R-1830 is the most produced aircraft engine of all time.

The P&W R-1830 Twin Wasp was one of the most reliable aircraft engines, perhaps second only to the R-2800. The R-1830 could handle a fair amount of abuse and continue to operate without issues. The engine was a favorite among pilots, and it typically only experienced issues when established practices were not followed. The R-1830 continues to be a reliable engine that powers many vintage aircraft.

P&W's first two-row radial was the experimental R-2270. Note that all the valves are driven from the front of the engine, resulting in 28 pushrod tubes leading from the cam ring to the cylinders. Subsequent two-row radials had separate valve actuation for the front and rear rows. (Photo Courtesy Aircraft Engine Historical Society)

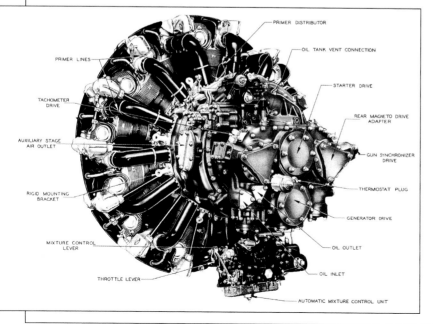

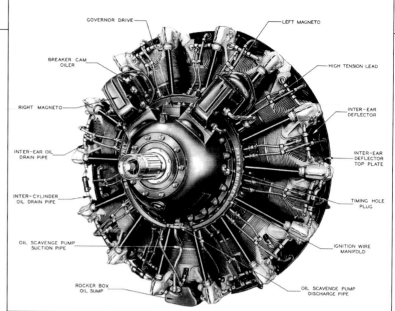

P&W's second two-row radial was the R-1830 Twin Wasp. The engine proved to be powerful and fairly compact, with a diameter of 48 inches. (Photo Courtesy Operators Handbook Twin Wasp C3 Engine, July 1942)

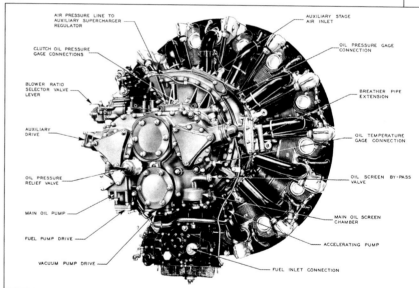

The R-1830 was an absolute success. It was in production from 1932 to 1951, and more R-1830 engines were made than any other aircraft engine. A total of 173,618 were built. (Photo Courtesy Operators Handbook Twin Wasp C3 Engine, July 1942)

The cylinder row stagger on the R-1830 was just enough to provide cooling air to the second row of cylinders and allow space for the intake and exhaust manifolds to reach the back of the front-row cylinders. (Photo Courtesy Operators Handbook Twin Wasp C3 Engine, July 1942)

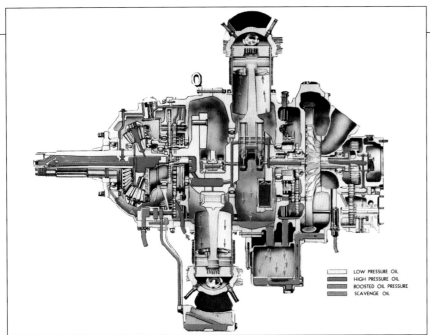

This cross section of the R-1830 illustrates the engine's oil system. Scavenge oil was taken from the sumps and passed through an oil cooler before it was returned to the oil tank. Note the minimal distance between the first and second cylinder rows to keep the engine as compact as possible. (*Photo Courtesy* Maintenance Manual Twin Wasp (R-1830) S1C3-G Engines, *December 1946*)

LOW PRESSURE OIL
HIGH PRESSURE OIL
BOOSTED OIL PRESSURE
SCAVENGE OIL

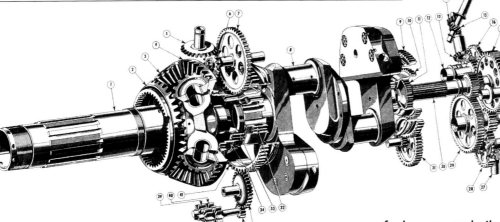

1) Propeller shaft, 2) fixed front gear, 3) gear reduction pinion, 4) rear driving gear, 5 and 6) governor drive gears, 7) front cam drive gear, 8) crankshaft, 9) rear cam ring, 10) rear crankshaft gear, 11) accessory drive adapter, 12) right magneto drive, 13 and 14) vacuum pump drive, 15) tach drive, 16) magneto, fuel pump, and oil pump drive gear, 17) accessory drive gear, 18) left magneto drive, 19) vacuum pump gear, 20, 21, and 22) oil pump idler, 23) oil pump drive shaft, 24, 25, and 26) oil pump, 27) vacuum pump drive, 28) generator drive gear, 29) impeller drive gear, 30) impeller shaft, 31) rear cam drive gear, 32) reduction drive gear, 33 and 34) reduction gear coupling, 35) oil pump drive, 36) front pump drive gear, 37 and 38) oil pump gear, 39 and 40) oil pump idler, and 41) front cam ring. (*Photo Courtesy* Instruction Manual Pratt & Whitney Engines R-1830-43 & 65, *November 1943*)

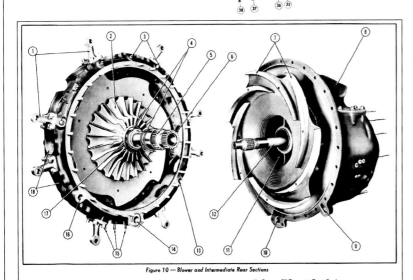

Figure 10 — Blower and Intermediate Rear Sections

1. Rigid Brackets
2. Impeller
3. Manifold Pressure Connections
4. Seal Rings
5. Impeller Shaft Low Ratio Gear
6. Impeller Shaft High Ratio Gear
7. Diffuser Vanes
8. Carburetor Mounting Flange
9. Return Oil Passage from Front Oil Pump
10. Scavenge Oil Passage from Sump to Rear Oil Pump
11. Intake Air Duct
12. Accessory Drive Shaft
13. Pressure Oil Passage to Front Crankcase
14. Scavenge Oil Passage from Sump to Rear Oil Pump
15. Sump Mounting Studs
16. Return Oil Passage from Front Oil Pump
17. Fuel Slinger
18. Intake Pipe Ports

In both P&W and Wright engines, the supercharger section was made up of two main components. The blower section, or front supercharger section (left), housed the impeller and distributed air to the cylinders. The rear section, or accessory section (right), contained the diffuser vanes for the impeller and also the drives for the supercharger and accessories. (*Photo Courtesy* Service Instructions R-1830-75 and -98, *February 1945*)

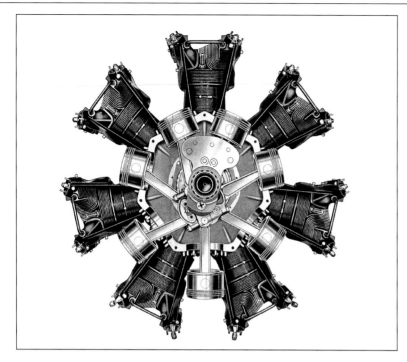

An unusual view of an R-1830 engine without its front cylinders, front crankcase section, and propeller reduction gear. The two-piece master connecting rod is visible on the right. Unless a special jig was used, the master rod cylinder was the first to go on and the last to come off the engine. Movement of the master rod beyond its normal operating positions could damage link rods and pistons. (Photo Courtesy Operators Handbook Twin Wasp C3 Engine, July 1942)

Here, a P&W R-1830 crankshaft undergoes a close visual inspection. The perfect machining on the crankshaft is all the more remarkable considering that the crankshaft was mass produced during wartime. Note the counterweights added to the cheeks of the single-piece crankshaft. (Photo Courtesy LOC 2017694125)

The R-1830 was mass produced by several manufacturers during World War II. Stacked in the background are center crankcase sections. In the foreground are front and rear sections that have been paired together. The rear crankcase section had a lip to which the supercharger section mounted. The front crankcase section had no lip. (Photo Courtesy LOC 2017694152)

Pratt & Whitney R-1830 Engines in Airliners				
Manufacturer	A/C Model	Engine Model	HP t/o	Number
Douglas	DC-3A	SB3-G	1,000	2
		SC-G	1,050	2
	DC-3C	S1C3-G	1,200	2
		S4C4-G	1,200	2
	C-47, A	R-1830-92	1,200	2
	C-47C	R-1830-90	1,200	2
Lockheed	14-08	S1C3-G	1,200	2
	18-08, -10,	S1C3-G	1,200	2
	18-14	S4C4-G	1,200	2
Martin	M-130	S2A5-G	950	4
Sikorsky	VS-44	S1C3-G	1,200	4

Major Applications of the P&W R-1830 Twin Wasp Engine	
Bristol	Beaufort torpedo bomber (various versions)
Budd	RB-1 Conestoga transport
Consolidated	B-24 Liberator bomber
	PBY Catalina seaplane
	PB2Y Coronado flying boat
	PB4Y-2 Privateer maritime patrol
Commonwealth	Boomerang fighter
Curtiss	P-36 Hawk fighter (various versions)
Douglas	DC-3 / C-47 airliner / transport (various versions)
	DB-7 Havoc I light bomber
	TBD Devastator torpedo bomber
FFVS	J 22 fighter
Grumman	F4F Wildcat fighter (various versions)
Lioré et Olivier	LeO 453 transport
Lockheed	Model 18 Lodestar transport
	Hudson recon bomber
Martin	M-130 flying boat airliner
	Model 167 Maryland light bomber
Republic	P-43 Lancer fighter / AT-12 Guardsman trainers
SAAB	B 17A light bomber / reconnaissance
	B/S 18A bomber / reconnaissance
Seversky	P-35 fighter
Shorts	Sunderland Mk V flying boat
Sikorsky	VS-44 flying boat airliner
VL	Myrsky fighter
Vultee	P-66 Vanguard fighter

(A) COWL MOUNT LUGS FRONT SUPPORT
(B) GOVERNOR
(C) FLEXIBLE COWL SUPPORT RING MOUNT
(D) COWL MOUNT STUDS – REAR SUPPORT
(E) SEAL BETWEEN BAFFLES AND COWLING
(F) VIBRATION ISOLATOR
(G) CARBURETOR PREHEAT HEAT EXCHANGER
(H) PREHEAT VALVE
(I) FIREWALL
(J) OIL TANK VENT LINE
(K) PREHEAT CONTROL
(L) MIXTURE CONTROL
(M) THROTTLE CONTROL
(N) BREATHER CONNECTION
(O) AUXILIARY DRIVE
(P) STARTER
(Q) MAGNETO
(R) MAGNETO BLAST TUBE CONNECTION
(S) OIL OUT TO OIL COOLER
(T) OIL IN FROM "Y" DRAIN
(U) GENERATOR
(V) GENERATOR BLAST TUBE
(W) FUEL LINE TO FUEL PUMP
(X) OIL DILUTION SOLENOID VALVE LINE
(Y) OIL RETURN LINE TO TANK
(Z) OIL OUT LINE FROM TANK
(AA) N.A.C.A. COWL
(BB) SEAL BETWEEN INNER AND OUTER DIAPHRAGM
(CC) ENGINE MOUNT RING
(DD) COOLING AIR EXIT FLAPS
(EE) STAMPED STAINLESS STEEL DIAPHRAGM
(FF) VACUUM PUMP
(GG) OIL COOLER
(HH) "Y" DRAIN
(II) OIL COOLER EXIT FLAP

Detailed here is what P&W considered a typical installation for an R-1830 engine. Every installation required certain components, such as an intake scoop and oil cooler; however, their design and location could vary depending on application. (Photo Courtesy Installation Handbook, March 1948)

Arriving at Douglas or any aircraft manufacturer, factory-fresh engines were put through a number of steps and inspections before they were mounted to the aircraft. It was the aircraft manufacturer's responsibility to mount the engine to the airframe and design an adequate cowling. (Photo Courtesy LOC 2017878897)

LANDPLANES BECOME VIABLE CONTENDERS

The airlines of the world were casting their views enviously at the flying boat technology of the 1930s, what with its lifting capability and range, none of it hindered by runway length. The land plane faction of commercial aviation was eager and trying valiantly to make a go of designing a capable heavyweight/long-range aircraft that could at least equal the impressive capabilities of the ocean-spanning airliners at Pan American Airways of the United States and Imperial Airways of England. Inch by inch and trial by exhaustive trial, the proper airplane made itself known, but not before a lot of pushing and prodding at the envelope's edges.

Getting There by Air

In 1946

With the new and luxurious, clean aerodynamic designs working for the airlines in 1935 and the

Clipper Flying Cloud *is the restored Boeing 307 Stratoliner on permanent display at the National Air and Space Museum's Udvar-Hazy Center at Dulles International Airport, Virginia. This is one rare airframe today. (Photo Courtesy Boeing Co. via Jon Proctor with permission)*

superb DC-3 just about to fly for the first time, the airline business was beginning to actually resemble a money-making enterprise. The country had traveled quite far from the mid-1920s to this point in terms of advanced aero equipment and its use: wing flaps, retractable landing gear, constant-speed props, radios, navigation equipment, and the emerging innovation of pressurized airplane cabins. The resultant efficiencies emboldened the airline management folks to ramp-up their wildest hopes for "super airliners." United Airlines led the way in autumn 1935.

United's boss was William "Pat" Patterson who shared the same acumen and daring as the presidents at the other four major airlines in this country at the time. He approached Douglas to request an airplane that would fly farther, faster, and higher than the DC-3 while carrying more passengers. Once again, the airplane was to have all the latest and greatest (and commensurately more complex) technology at hand. This necessitated the use of four engines and a pressurized cabin. The Santa Monica firm responded with reticence, but when Patterson informed Douglas that the other four airlines would each participate in the venture and buy 20 airplanes each, Douglas decided to go ahead with the project. Thus, the one-of-a-kind DC-4E ("E" for Experimental), was born.

The airplane evolved into a 66,000-pound airliner capable of flying 2,200 miles at 200 mph while carrying either 32 or 42 passengers in nighttime configuration (berths again) or 52 in a daytime layout. The airplane was large, and its fuselage diameter was wide enough to handle two rows of two seats throughout the length of the cabin. Because the airlines wanted a tricycle landing gear arrangement, the single tail would not fit into existing hangars and so, the airplane received three smaller vertical stabilizers installed on its horizontal tail plane. Cruising altitude with pressurization was set at almost 23,000 feet. All this was a mere 10 years after the Ford Trimotor. Both literally and figuratively, the DC-4E was a legitimately big deal.

The interior of the aircraft was laid out such that the aft men's dressing room had three wash basins, a sofa, and an enclosed toilet. Next to it was the electric galley kitchen. Two stewardesses would manage the large number of passengers, and they had their own seating near the aft cabin door, behind a partition. Farther aft, the designers had even conceived a separate Bridal Suite with its own facilities, and the ladies lounge and bathroom was next door to that. The airplane, flown by a crew of three, seemed immense. Upon rollout from the hangar, it actually broke through the concrete of the ramp area.

First flying in June 1938, the DC-4E proved early on that although a good flying airplane, it was indeed complex. The engines gave fits. Once operated by United in mock route trials, the airline quickly discovered just how expensive a large and very complicated airliner was to operate and maintain. TWA and PAA had opted out of the experiment, and the remaining American and Eastern decided to rescind their options to fly the airplane over their existing routes. The whole program, although a financial loss for Douglas, was an integral cog in the wheel of progress toward large land-based airliners.

By the late 1930s, "big" was in as it referred to airplane design. Perspective allows you to place the DC-4E and upcoming designs into context by mentioning the following airplanes: XB-15, XB-19, B-17, and 307 Stratoliner. At Boeing, the XB-15 experimental heavy bomber of 1936 was, at the time, the world's largest and heaviest airplane. Also at this same time, design studies were underway at Douglas for the XB-19 Hemisphere Bomber, larger yet than the XB-15. Also at Boeing dating from 1935 was its model 299, the four-engine design for a bomber competition held by the Army Air Corps. Losing that particular first contract, this sterling airplane went on to become the inimitable B-17 bomber of World War II. It was the direct antecedent of the Boeing Model 307 Stratoliner.

Boeing had a penchant for decades, before and after the war, to use its military designs as springboards for its commercial airplanes.

The 307 was a shining example of this philosophy because it consisted of the vertical and horizontal stabilizers, landing gear, wings and engines/nacelles, and basic systems of the company's B-17. Only the fuselage was different, made large enough to seat 33 passengers and carry a crew of 5. The range of this airliner was just shy of 2,400 miles, which landed it in the transcontinental category. But the big news for this big airplane was that it would be the first operational production airliner in the United States to be pressurized. This was a breakthrough.

With a service ceiling of 26,000 feet, the Stratoliner (aptly named) would now give its fares a smooth ride more of the time by cruising above most weather disturbances. PAA and TWA were the launch customers, KLM having also flown the airplane but withdrawing its interest. The first flight for the 307 was the last day of the year 1938. By 1940, it was in service with both carriers. The cost per plane was $315,000, a spectacular amount late in the decade. PAA flew three airframes and TWA, six.

Often, one's reach exceeds one's grasp, and although dreams can be gargantuan, the reality of the final product is more realistically muted, tempered by all the real-world constraints. The 307 was a perfect example of this diminution. The DC-4E was the big dream, the Stratoliner the refined, practical end result that was necessarily of smaller size. Nevertheless, the Boeing airplane was a superbly luxurious airliner for its passengers. It was spacious, smooth, capable of sleeping accommodations, and state of the art for its day. The galley kitchen had come into its own, allowing for hot meals and drink, and bathrooms and facilities were more open and enjoyable, taking the great start of the DC-3 to new levels. And of course, we mustn't ever diminish the impact that pressurized flying had on air travel. Diversions and cancellations dropped significantly thanks to surmounting of the weather en route.

So, look at where we were in that first year of the tumultuous 1940s: Boeing 314 flying boats spanning the world's oceans and Boeing 307 land planes crossing the United States and Latin America; both of these aircraft were highly innovative and unbelievably large. The horizons were limitless for commercial aviation and the airlines. Unfortunately, at precisely that moment a world war was thrust upon air travel aspirations and all progress was diverted to military applications. The good and bad news for commercial aviation and aeronautics was that things would never again be the same as those golden years of the 1930s.

Come 1942, Douglas had learned its lesson concerning practicality in an airframe design. The redux of the DC-4 had begun, and in February of that year, the first flight of the new, smaller, single-tail, and unpressurized four-engine airliner was at hand. The DC-4 had been officially drafted by the Army Air Force and designated C-54

The DC-4E was essentially an expansion of the DST/DC-3 planform and design philosophy, from its rectangular windows to the small berth windows in the upper fuselage and shape of the nose and windscreen. Note that the engine nacelles are rather perpendicular to the swept leading edge of the wing, and not aligned with the aircraft's line of flight. (Photo Courtesy Museum of Flying Collection)

As the only airline still involved with the DC-4E, United used it extensively on route-proving flights and to gauge public opinion. The relatively immense size of the airplane is obvious at this open house in Denver, Colorado.

Comfortably flying with two engines shut down, the grandeur of this design is readily apparent. As with many other aviation firsts, the DC-4E was just too much too soon. (Photo Courtesy Museum of Flying Collection)

With its immense wing-chord dimensions, the XB-15 experimental bomber was, like the DC-4E, a stretch of the envelope. However, Boeing used the invaluable data gleaned from this airplane in a scaled-down version (the B-17) and a scaled-up version (the B-29), and helped win a war in the process. At Douglas, DC-4E knowledge led to a smaller, more advanced version as well: the C-54 transport, which contributed significantly to allied victory in World War II and became the production DC-4 airliner.

The gargantuan nature of the Douglas XB-19 Hemisphere Bomber cannot be overstated. Each progressive step of aerodynamic growth in the period between the wars led to tangible applications further down the road. The maximum range of the B-19 was a whopping 7,200 miles, but it only cruised at a pedestrian 135 knots. (Photo Courtesy Museum of Flying Collection)

Skymaster. It became a world beater, airlifting and supplying US forces literally around the world, particularly in the long-span areas of the Pacific Theater. Douglas Aircraft's growing reputation for efficient and reliable transports was completed by the Skymaster. The new-and-improved DC-4 therefore became the foundation for the world's airlines immediately following the end of the conflict in 1945.

Douglas had produced 1,165 C-54 variants for the army and the navy. As peace returned to the country and the world, the manufacturer offered the civilian DC-4 in its new guise as the immediate answer to the airlines' lift requirements. The airplane was in production at both Santa Monica and Chicago and could easily be pressurized and demilitarized. When compared to the mainstay DC-3, the big brother of the line was faster by 50 mph, could carry 44 passengers, and could fly three times as far as the twin. This was obviously a quantum leap in size and capabilities. The airlines were ready to order and Douglas ready to deliver. United started the parade, and Northwest, Western, National, Air France, Scandinavian Airlines System (SAS), and Sabena followed.

Then the US government pulled the rug out from underneath the program. It made available the majority of war-weary C-54 airframes, and at not-unexpected bargain prices. A new production DC-4 from the line was priced from $385,000 to $595,000, while the most expensive of the military Skymasters was a mere $100,000. The effect was swift and obvious. Only 79 "real" DC-4s were built for the airlines, the rest (and there were plenty) consisted of war-surplus examples that had been converted either by outside vendors or the airlines themselves. This is why so many photos of DC-4s show the cargo doors at the rear of the fuselage, even while wearing airline liveries.

American Airlines was also one of the convertees, but it insisted on replacing those cargo doors with a single-passenger entrance. Regardless, any flight on a DC-4 was spacious and a real boost from the taildragger era of DC-2s and DC-3s. The traveling public was now beginning to enjoy the true comprehensive benefits of flying across the country and doing so in a modern airplane. We had come a long, long way from Fokker 10s.

Not only is this a rare airplane, but seeing a PAA Boeing 307 in postwar colors is just as rare! A typical day at work flying from Havana, Cuba, to Miami. Notice the air-conditioning truck with its hose connected to the aft fuselage of the airplane. (Photo Courtesy Boeing Co. via Jon Proctor, used with permission)

Departing westbound from Miami International airport, the wide track of the 307's main gear is apparent here. PAA had only three Stratoliners in its fleet. (Photo Courtesy Gerald Balzer Collection)

This extremely rare shot of the Boeing 307A in color is TWA's highly polished NC19905 at Chicago's Midway airport on 10 October 1941. Notice the lineup of DC-3s behind the Stratoliner. Less than two months later, America was at war and the 307 was conscripted to military service, its luxury interior stripped. (Photo Courtesy Charles W. Cushman Collection, Indiana University)

Artist's rendering of the TWA 307's interior. What is striking about the 307 was its compartmentalized use of space. The airplane was not just a tube with seats in it, but rather more like a flying boat, cruise liner, or even one's living room.

This promotional package sent to regular TWA customers included a multipage color spread in Collier's magazine and postcards filled with corporate copy extolling the virtues of the airplane. Make no mistake about the significance of this aircraft in passenger service: The first pressurized cabin in the United States was a very big deal indeed, a fact not lost on the average passenger of the time

Broader Wings

... AN THE NATION ... And the graceful sweep of their arrival ... TWA's new Boeing Stratoliners are here! And with them come ... never before known along any commercial airway in the ... again, as in the past, is first to put the proved into pr... oes hand in hand with Precision on the airl... NENTAL & WESTERN AIR

TWA Colorlado

TWA Stratoliner in Flight

KANSAS CITY MO.

olthoff Western R a, Ill.

CROWELL-COLLIER PUBLISHING COMPANY—PUBLISHERS OF COLLIER'S—THE AMERICAN MAGAZINE—WOMAN'S HOME COMPANION

June 1, 1940

5¢ A COPY

Collier's
THE NATIONAL WEEKLY

There was lots of room in the large galley to prepare the trays. (Photo Courtesy Jon Proctor)

We are indeed fortunate to have color photos of TWA's 307 interior. The size and sumptuousness of the club seating on the starboard side of the cabin is almost overwhelming. Realize that this arrangement borrowed directly from Boeing's 314 flying boat. Even the upholstery is the same. What wouldn't we give today to fly across the country in this kind of superb luxury? (Photo Courtesy Jon Proctor)

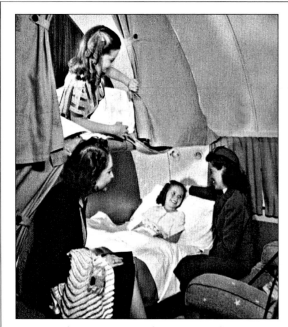

As was the custom in the 1930s, sleeping aboard your airliner was expected, owing to the long hours required to cross the United States by air. Because these were erected above, rather than pull-down berths, you wonder where this contraption was stowed when not in use. (Photo Courtesy Jon Proctor)

When the war ended, 307s came home to both PAA and TWA. They were modified to incorporate refinements from the B-17G bombers, which included new horizontal stabilizers, landing gear, wing panels, and nacelles. Unfortunately for the passenger, the superb interior furnishings were not replaced, and the airplane reverted to the industry standard "flying tube with seats." TWA's updated markings were also applied to the airplane, experimenting with white edges to enhance the red stripes on bare metal. Following this, TWA reverted to black pinstriping on its fleet. (Photo Courtesy Jon Proctor)

These DC-4s on the pre-delivery ramp at Santa Monica were destined for the following airlines: Air France, Delta, Mexicana, Swedish, United, and Western. These were manufactured as "pure" DC-4s, not conversions from C-54s. Note the Brazilian government DC-3D, and in the background a very rare B-23 Dragon bomber. (Photo Courtesy Museum of Flying Collection)

Keeping the chief clean, a candid shot inside the LAX hangar for Western Air Lines. It is no coincidence that the man with a hose has coveralls that are indeed covered with oil, the lifeblood of radial engine aviation.

Upper Left: One of the newly manufactured DC-4s flown by Western over the San Francisco area. (Photo Courtesy Jon Proctor)

PAA's DC-4 over the San Francisco Bay Bridge. DC-4s, along with the original Constellation, made flying boats obsolete for the airlines, particularly PAA. All that extensive globe-girdling military travel on such a routine basis during World War II indirectly delivered the airline business into the accepted international conveyance it has since become, with "land planes" assuming the preeminent role within air transportation. (Photo Courtesy Gerald Balzer Collection)

No survey of airliners would be complete without mention of the lost member of the Douglas family, the DC-5. Meant as a DC-3 replacement in the 1940s and designed by the military aircraft division of Douglas, the airplane didn't quite measure up, with only KLM operating the type in revenue service and the US Navy and Marine Corps operating the aircraft during prewar and wartime periods. William Boeing owned one example for his private use. Now there was a statement of approval! (Photo Courtesy Gerald Balzer Collection)

A shiny new Delta DC-4 conversion at the factory. The prewar paint scheme adorns this airplane, but not for much longer as the DC-6 arrived. Delta's airplanes carried 44 passengers and competed against Eastern's pressurized Constellations. (Photo Courtesy Jon Proctor)

One of American's C-54 conversions executed under contract by Republic Aviation Corporation on Long Island. How ironic that the advanced 400-mph Rainbow (see chapter 5) was not produced for American Airlines and that Republic instead ended up doing engineering rework for American's ponderous DC-4 fleet.

Consolidated Vultee (Convair) took a novel approach to postwar cargo lift and devised the Model 39 Liberator Express, based on the PB4Y and B-24 bomber. American had determined that air freight was the way to go and for 20-plus years maintained dedicated cargo aircraft in its fleet. Although the Express was tested thoroughly for several months by hauling all types of cargo, it was finally determined to be underwhelming compared to surplus C-54s available to do the same job. (Photo Courtesy Gerald Balzer Collection)

Turning a Corner

It is the wise man who does not sneer at a diminutive, hungry, and therefore dynamic and determined small company that desperately wants to break into the Big Time and make its mark on the world. Such an enterprise was again found in Southern California and went by the name Lockheed Aircraft Corporation.

Having given the airline world some very successful smallish twin-engine all-metal airliners, Lockheed decided to up the ante and enter the four-engine arena with Douglas and Boeing in 1939. It wooed PAA into working on a speedy 44-passenger airliner known as the Excalibur. This concept grew yet again to finally come to rest as the now-familiar Constellation triple-tailed airliner.

Subordinated to the army due to the encroaching war, the Connie, as it became known, was designated as the C-69 and made its first flight one year after the C-54 in January 1943. Lockheed's model 049 was one impressive airplane. Here are some stats: speed of up to 329 mph, finally breaking the 300 barrier, and carrying 51 passengers nearly 3,000 miles with a fuselage at its widest point 2 feet wider than the DC-4. The airplane was pressurized from its inception. Both TWA and PAA were onboard as its first customers after the military. The C-69 had some teething problems and saw limited service with the army in those first two years. Its real flying began just after the war.

Following V-J Day, the military canceled the remaining C-69s. Lockheed made the gutsy decision to keep the Connie production lines open, buy back the military C-69s, and finish any uncompleted airframes as civilian Model 049s. This allowed the company to steal a march on Douglas with its pressurized, fast DC-6, which was not due for another two years.

Being first can be profitable. It can also be premature. In the case of the Connie, there were plenty of systems problems to surmount on the way to dependable operation as an airliner. One notable hiccup was the heating and cooling system. Inadequate for the job, the airplane ran temperature zones wherein parts of the cabin were too cold, others too hot. There was never consistency. It even got to the point that individual, rotating, caged table-top–style mini fans that hung from the coat rack above the seats were installed.

No matter though, the Constellation was the only pressurized airliner you could fly aboard in the immediate postwar period, and don't you know TWA and Eastern had something to boast about with that factoid. TWA began New York to Paris service with its Constellations in February 1946, and Los Angeles to New York service in March.

TWA inserted a clause in its contract with Lockheed that was the karmic replication of the United-Boeing agreement with the 247. It stated that no other airline could order and operate the Constellation on west-to-east routes (meaning transcontinental) for the first

The first C-69 wings eastward to Washington, DC, for presentation to the USAAF. As part of a clever promotional campaign devised by Howard Hughes, then-owner of TWA, airline markings were painted on the airplane to inform the world that TWA would be the first domestic airline to operate the new, fast, and pressurized airliner after the war. The military brass were not amused. (Photo Courtesy Jon Proctor)

One of only 15 Lockheed C-69 aircraft for the USAAF standing on the ramp at Burbank, California. To contrast its size, the company's design effort for the postwar general aviation market stands nestled underneath. That light airplane is the Big Dipper (remember that all Lockheed airplanes were named after stars in the heavens) and you can date the photo by noting that the Dipper first flew in July 1945 and crashed on 6 February 1946. Eventually, all C-69s were modified by Lockheed and resold to the airlines.

049 Constellation for ultra-early customer American Overseas Airlines (AOA). Here the airplane is parked at LaGuardia Airport, New York. AOA was the third airline to receive Constellations and flew seven airframes until 1950 when PAA acquired the division from American Airlines. This is NC90922, Flagship Copenhagen.

Modern airliner cockpit in 1945. Compare this collection of crowded instrumentation and levers, knobs, and switches to the Ford Trimotor in chapter 1. How far cockpit design had progressed, and yet, note the two small trays with cheesecloth over them atop the glare shield. I flew with a pilot who had many hours aboard early Constellations, and who made the point that these were reservoirs for alcohol that would be ignited by the crew with matches to defrost the windshield!

airline business, and each successive step forward allowed one carrier to up the ante for the entire industry. Indeed, airplane order books were based upon this "firstest with the mostest" mentality, with the next biggest and best "thing" driving the tone and tenor of design shops everywhere.

In 1945–1946, it became possible to completely cross the United States, coast to coast, in a little more than the number of hours in a business day. On the Connie, it was 9 hours 15 minutes eastbound. Yes, there was a stop in the middle of the country (Kansas City for TWA, Chicago for United, or Fort Worth for American), but nevertheless, with the new Constellation and its slower counterpart the DC-4, businessmen were now unshackled to move about the country unimpeded and build the new postwar society. Air transportation was finally legitimate.

two years after TWA took delivery of the airplane. PAA was alright, owing to its international use, as was American Overseas Airlines (AOA), and certainly Eastern was fine by flying its north to south East Coast routings.

Product differentiation was always the name of the game in the

This is one of PAA's 049s, delivered in September 1946, Clipper Southern Cross, later Hotspur. Engines are being run-up over at the maintenance facility on the north side of San Francisco International airport. (Photo Courtesy SFO Museum Collection)

The temperamental Wright R-3350 engine powered the Constellation and the Boeing B-29. Here, the exhaust collector ring is being attached to the cylinders. (Photo Courtesy Gerald Balzer Collection)

Wright was proud to be a part of the burgeoning international aviation trend with the Constellation at its heart.

Pearce

Hardware on the Wing

Pratt & Whitney R-2180 Twin Hornet

In 1933, P&W aspired to build an engine that was larger and more powerful than the R-1830. This new, two-row radial design was known as the R-2180 Twin Hornet. The engine had two rows of seven cylinders. Contrary to the Twin Hornet's name, its cylinders were actually based on those of the Wasp engine. The R-2180 had a 5.75-inch bore, 6.0-inch stroke, and total

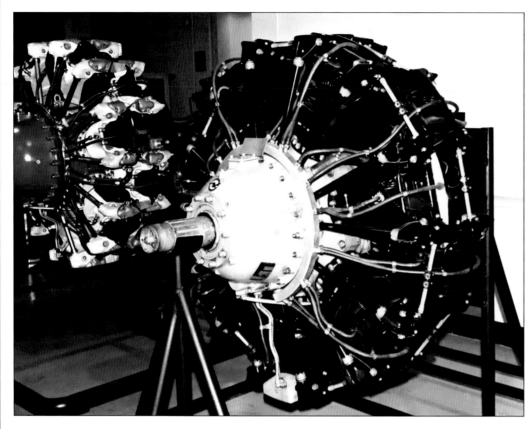

displacement of 2,181 ci. The 14-cylinder engine had a 6.66:1 compression ratio and used a 0.50 propeller gear reduction via bevel gears. The engine had a diameter of 51.61 inches and weighed 1,675 pounds. The R-2180 produced 1,400 hp at 2,500 rpm for takeoff and 1,150 hp at 2,350 rpm for normal operation.

The R-2180 was first run in 1936 and performed well. The engine's design continued the use of two-piece master connecting rods and a one-piece crankshaft. Just as the R-2180 entered limited production in 1937, P&W became aware of the Wright R-2600 and R-3350 engines. Realizing the Twin Hornet was eclipsed, and also having other serious engine projects in the design phase, P&W decided to terminate development of the R-2180 engine. Only 30 R-2180 engines were made. An upgraded version with a two-speed supercharger and single-piece master rods was planned but never built.

The resources freed up by the R-2180's cancellation were directed to P&W's next engine, the R-2800, and some R-2180 components were later used on the R-4360 prototype. The R-2180 was an important step toward a better engine, the R-2800. The only airliner powered by the engine was the experimental Douglas DC-4E. Although just one DC-4E was built, experience with the aircraft ultimately led Douglas to create the DC-4. The R-2180 engine was later redesignated R-2180A when a new engine, the R-2180E Twin Wasp, was introduced in the late 1940s.

The Douglas DC-4E was originally known as the DC-4. When Douglas and the airlines realized the aircraft was too large and complex to enter airline service, the "E" was added to signify "Experimental." The DC-4E and the Stearman XA-21 were the only aircraft flown with P&W R-2180 engines. The DC-4E was sold to Japan and later served as the basis for the Nakajima G5N Shinzan heavy bomber. (Photo Courtesy William Pearce Collection)

Pratt & Whitney R-2180 Engines in Airliners				
Manufacturer	A/C Model	Engine Model	HP t/o	Number
Douglas	DC-4E	S1A-G	1,400	4

Wright R-2600 Twin Cyclone

Wright began developing two-row radials around 1931. The company's first pair of two-row engines were extensions of the Whirlwind series, and each had 14 cylinders. The first was the R-1510, which had a 5.0-inch bore and 5.5-inch stroke. The engine displaced 1,512 ci and produced 670 hp at 2,400 rpm. The R-1510 weighed 1,025 pounds and was first run in 1932. The second two-row engine was the R-1670. Its bore was 5.25 inches, a slight increase from the R-1510, but the stroke remained at 5.5 inches. The R-1670 displaced 1,667 ci and produced 850 hp at 2,600 rpm. The engine weighed 1,236 pounds and was first run in 1934. Only 34 R-1510 and one or two R-1670 Twin Whirlwinds were made, and the engines did not enter production. However, Wright gained experience that was applied to its next, larger two-row radial.

In 1935, Boeing approached Wright requesting a 1,500-hp engine for the new flying boat that the company was designing, which became the 314 Clipper. Wright requested engine proposals from 12 of its top designers. Encouraged by coworkers, Rudolph (Rudy) Daub, a draftsman originally from Germany, submitted an unsolicited design for a two-row 14-cylinder radial engine. On 29 November 1935, Daub was notified that his unsolicited design had won and that he was now the engine project's head designer. This new engine surpassed P&W's R-1830 in both engine displacement and power.

On 1 December 1935, official design work on Daub's engine was started. The design was loosely based on the R-1820 Cyclone. The new engine had 14 cylinders with the same 6.125-inch bore as the Cyclone, but the stroke was decreased from 6.875 to 6.3125 inches. Daub's engine displaced 2,604 ci and had a 55-inch diameter. It was known as the R-2600 Twin Cyclone (or Cyclone 14).

The R-2600 was constructed in a similar fashion as the P&W R-1830. The forged aluminum crankcase was made up of three

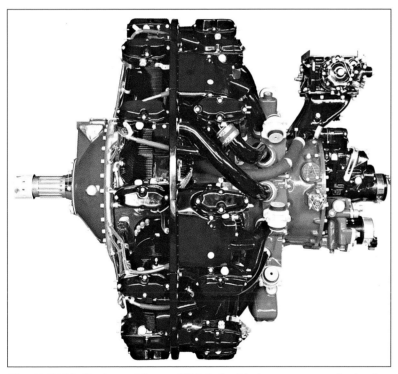

The R-2600 Cyclone 14 was Wright's third attempt at designing a successful two-row radial engine. Although some difficulties were encountered, the engine proved to be reliable. When viewed from the rear, the intake port for front-row cylinders is on the right, and the intake port for rear-row cylinders is on the left. (Photo Courtesy Aircraft Engine Historical Society)

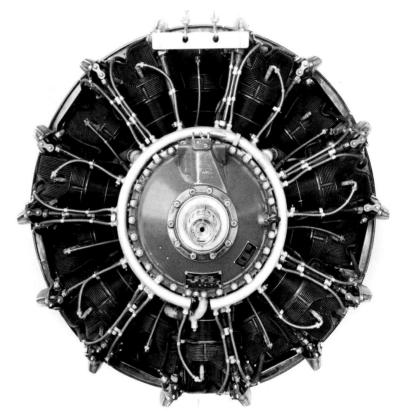

P&W had nothing that could compare to Wright's R-2600 when the engine first ran. The R-2600 would probably have powered more airliners than the Boeing 314 had World War II not intervened. (Photo Courtesy Aircraft Engine Historical Society)

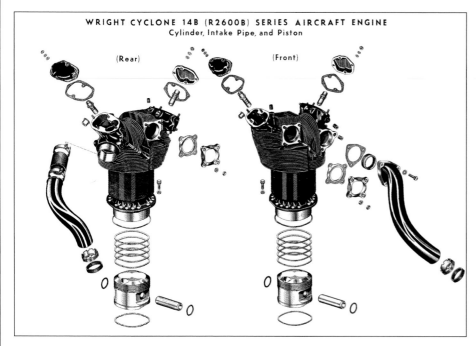

WRIGHT CYCLONE 14B (R2600B) SERIES AIRCRAFT ENGINE
Cylinder, Intake Pipe, and Piston

(Rear)   (Front)

The rear and front cylinders for the R-2600 had different port placements to enable manifolds to bring in the fresh air and fuel charge and carry away exhaust gases. The ports with square flanges were for exhaust. (Photo Courtesy Parts Reference Charts Wright Aircraft Engines Cyclone 14BA, October 1944)

sections split vertically through the cylinders. In later models, stronger crankcase sections were made of thin-wall steel forgings that reduced crankcase thickness by about half an inch. Although the steel crankcase strengthened the engine, it also added weight. The engine's nose case and rear section were both made of magnesium to save weight. The nose section housed planetary gears for the propeller gear reduction, and common ratios were 0.5625 and 0.4375. The rear section housed the supercharger and drove accessories. Initially, a Wright-designed single-speed supercharger was used, but later models had a two-speed supercharger.

One difference from the P&W R-1830 was that Wright used a three-piece crankshaft and single-piece master connecting rods. The master rods for the R-2600 were located in cylinders number-1 and -8. In typical Wright fashion, the crankshaft was connected via clamping collars at the crankpins. Wright also used copper-lead bearings attaching the master rod to the crankpin but later switched to the silver-lead-indium bearings developed by P&W.

Initially, the R-2600's basic cylinder design was from the R-1820 Cyclone with a steel barrel shrunk and screwed into a cast-aluminum head with a hemispherical combustion chamber. As the R-2600 was developed, the machined cooling fins on the cylinder's steel barrel were replaced by a series of small grooves. An aluminum muff made up of Wright's thin, high-strength W cooling fins was rolled into the grooves in the steel barrel. The use of the aluminum muff increased cooling fin area,

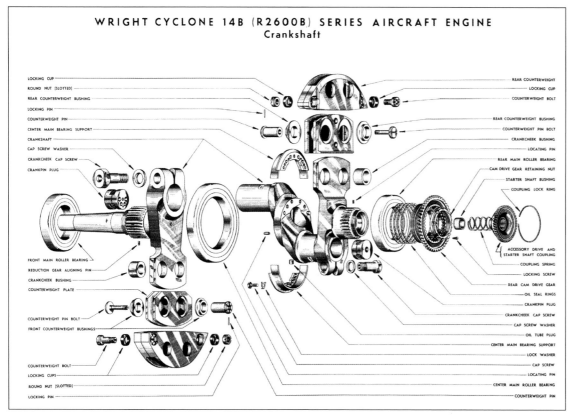

WRIGHT CYCLONE 14B (R2600B) SERIES AIRCRAFT ENGINE
Crankshaft

The R-2600's crankshaft was a continuation of Wright's successful clamping-style design. This drawing shows the front crankshaft section free; the rear crankshaft section has been attached to the center crankshaft section. (Photo Courtesy Parts Reference Charts Wright Aircraft Engines Cyclone 14BA, October 1944)

The Boeing 314 Clipper was essentially the last large flying boat airliner made in the United States. Its R-2600 engines were housed in NACA-style cowlings with hatches that allowed access to various accessories. Note the exhaust stacks on top of the cowlings. The 314 entered service right before World War II and was the only airline application for the R-2600 engine. (Photo Courtesy NASM SI-85-142-39)

in November 1936, and passed its type test on 10 June 1937. However, putting the engine into quantity production took some time to sort out. The delay led some aircraft manufacturers to switch to the more-powerful P&W R-2800, which was developed after the R-2600 and was just becoming available. Prototypes of the Curtiss C-46 Commando transport and the Grumman F6F Hellcat naval fighter were built and flown with the Wright R-2600, but production versions of the aircraft were switched to the P&W R-2800.

To make matters worse, quantity production of the engine was rushed due to the impending war in Europe and the expansion of the United States Armed Forces. Wright was busy producing other engines and expanding production, leaving development of the R-2600 behind what it should have been. Carburetor issues, excessive oil consumption, and rust and corrosion in the cylinders were all problems that needed to be overcome. The rust issue became so prevalent that in 1943, the FBI investigated for possible sabotage; the investigation did not uncover any foul play. A rust preventive and a modified Bakelite varnish sprayed and baked on all steel parts solved the problem.

The R-2600 A-series entered limited production around February 1937. The A-series initially had a 6.3:1 compression ratio and was rated for 1,500 hp at 2,300 rpm for takeoff and 1,125 hp at 2,100

and it was replaceable if damaged. One cam ring was positioned at the front of the engine, and another cam ring was at the rear of the engine. Each of the cam ring's two tracks had three lobes, and the ring was driven at 1/6 crankshaft speed in the opposite direction of the crankshaft. Via roller tappets and pushrods, the cam rings actuated the one intake valve and the one sodium-cooled exhaust valve in each cylinder. Ports for the intake and exhaust were on the back side of all the cylinders.

Initial development of the R-2600 went smoothly. The engine was first run in June 1936, completed a 150-hour endurance run

At the North American Aviation plant, R-2600 engines were attached to mounts and cowled before being installed in B-25 Mitchell medium bombers. Studebaker was selected to produce the R-2600 engine, but Wright objected. As a result, Studebaker built R-1820 engines, and Wright produced all R-2600s made. (Photo Courtesy LOC 2017878511)

rpm for normal operation using 90-octane fuel. Development of the R-2600 led to the B-series, which entered production around March 1940. B-series engines had a 6.9:1 compression ratio and used 100-octane fuel. Changes to the engine brought power for the B-series up to 1,700 hp at 2,600 rpm for takeoff and 1,500 hp at 2,400 rpm for normal operation.

The R-2600 C-series entered production around February 1943. These engines had a steel crankcase and a two-speed supercharger. A steel crankcase was part of Daub's original R-2600 design, but he was unable to convince Wright management to incorporate it. C-series engines had a takeoff rating of 1,900 hp at 2,800 rpm and a normal rating of 1,600 rpm at 2,400 rpm using 100/130 PN fuel. The engine's weight changed over its production life, from 1,935 pounds for the A-series to 2,045 pounds for the C-series. The R-2600 saw widespread service in World War II, but Wright was the sole producer of the engine.

A plant was built at Lockland, Ohio, for R-2600 production. With the rapid expanse of war-time production, Wright did not have the management in place to properly supervise the Lockland plant. The plant suffered from inattentive labor and the use of improper practices and faulty materials. The outcome of these issues was that defective engines were shipped and subsequently installed in aircraft, leading to a rise in engine failures. The plant production problems led to an investigation by the Truman Committee in 1943. The investigation revealed that Wright management at Lockland had essentially bribed civilian and government inspectors to pass the substandard engines.

As a result of the investigation, some personnel were dismissed, others were reassigned, and a few were charged with negligence. More stringent inspections were implemented, which led to an increased rejection rate and a subsequent drop in production. For example, of the 1,700 engines scheduled for May 1943, only 829 were completed. This production shortfall was an exaggeration of the actual number of defective engines because even the most minor, inconsequential issues led to a rejection. In fact, some studies showed that the failure rate of improperly inspected engines was not significantly higher than the failure rate of engines produced under the improved inspections. Eventually, the issues were sorted, and full production resumed, but Wright's credibility suffered as a result of the investigation.

Rudy Daub had always warned of Hitler's Germany, but because of close ties with some of his countrymen, Daub was prohibited from entering any Wright Aeronautical facilities and working on any projects during the war. One can only speculate how development of the R-2600 and R-3350 could have been improved if the man who designed the engines had been allowed to continue his

work. Production of the R-2600 ended in 1946 with a total of 85,374 engines built.

After a bit of a troubled start, the Wright R-2600 Cyclone 14 proved to be a reliable engine that gave good service, particularly in North American B-25 Mitchell medium bombers. The engine had very limited service in airliners, only being applied to the 12 Boeing 314 Clippers that were built. However, the four R-2600 engines installed in Clipper 18602 operated with little fault on one incredible journey that the aircraft was forced to make.

Caught in New Zealand at the start of World War II, Clipper 18602 made a 22,000-plus-mile, month-long journey around the world from Auckland to New York via Africa. The flight stressed the engines with over-gross takeoffs and the use of 90-octane fuel on some legs because 100-octane fuel was not available in remote areas. On the morning of 6 January 1942, Captain Robert Ford brought Clipper 18602 in for an unscheduled landing at the Marine Air Terminal at LaGuardia, much to the surprise of the controllers.

Wright R-2600 Engines in Airliners				
Manufacturer	A/C Model	Engine Model	HP t/o	Number
Boeing	314	GR260A2	1,500	4
		GR2600A2A	1,600	4

Major Applications of the Wright R-2600 Cyclone 14 Engine	
Brewster	SB2A Buccaneer scout bomber
Boeing	Model 314 Clipper flying boat airliner
Curtiss	SB2C Helldiver dive bomber
Douglas	A-20 Havoc attack bomber
	B-23 Dragon medium bomber
Grumman	TBF (TBM) Avenger torpedo bomber
Martin	Model 187 Baltimore light bomber
	PBM Mariner flying boat
Miles	M.33 Monitor target tug
North American	B-25 Mitchell medium bomber
Vultee	A-31 / A-35 Vengeance dive bomber

Pratt & Whitney R-2000 Twin Wasp

In January 1939, Douglas approached P&W in need of a new aircraft engine. Douglas planned to salvage what it could from the development of the DC-4E and create a smaller aircraft, the DC-4. The DC-4E was originally powered by P&W R-2180 engines, which were discontinued. Douglas needed engines more powerful than the R-1830 but smaller than the R-2800.

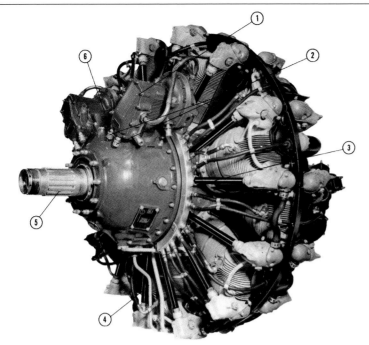

The R-2000 was based on and very similar to the R-1830. 1) Magneto, 2) magneto ventilators, 3) ignition manifold, 4) front oil pump suction tube, 5) propeller shaft, and 6) propeller governor drive. (Photo Courtesy Handbook Overhaul Instructions R-2000-4 Aircraft Engines, May 1957)

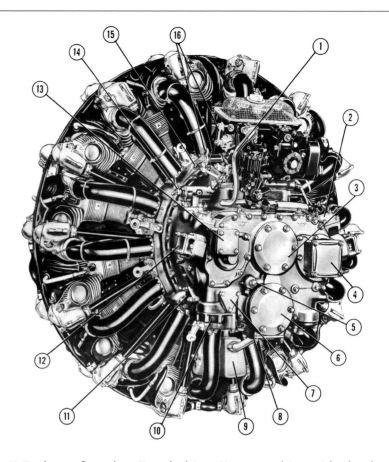

1) Fuel transfer tube, 2) tach drive, 3) starter drive, 4) hydraulic or vacuum pump drive, 5) oil pressure relief valve, 6) generator drive, 7) oil outlet, 8) vacuum pump drive, 9) oil sump, 10) rear oil pump, 11) oil inlet, 12) left auxiliary drive, 13) fuel pump drive, 14) selector valve, 15) manifold pressure connection, and 16) carburetor controls. (Photo Courtesy Handbook Overhaul Instructions R-2000-4 Aircraft Engines, May 1957)

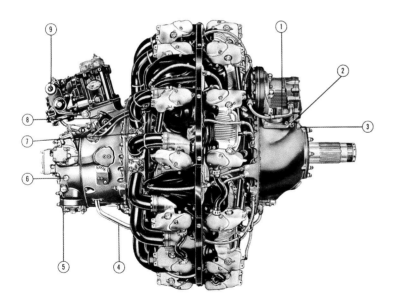

The R-2000 had a 0.25-inch larger bore than the R-1830. 1) Magneto high-tension lead, 2) magneto ground terminal, 3) magneto ventilators, 4) oil sump vent tube, 5) bypass valve, 6) oil inlet thermometer connection, 7) engine mounting bracket, 8) accelerating pump, and 9) fuel inlet connection. (Photo Courtesy Handbook Overhaul Instructions R-2000-4 Aircraft Engines, May 1957)

P&W found a simple solution to the Douglas request. The bore of the 14-cylinder R-1830 engine was enlarged by 0.25 inch to create a new engine, the R-2000. With its 5.75-inch bore and 5.5-inch stroke, the R-2000 displaced 1,999 ci. The R-2000 retained the Twin Wasp name, and R-2000 engines carried a continuation of R-1830 serial numbers. In fact, the R-2000 engines were considered the D-series of the R-1830 Twin Wasp line.

Early R-2000 engines used the same crankcase as the R-1830. However, the enlarged openings for the cylinders and the engine's extra power proved to be too much for the crankcase, and cracks resulted. A new crankcase was designed, and other changes were incorporated. The R-2000's crankshaft bearings were of the plain type, not the roller bearings used on the R-1830. The two-piece master rods were located in cylinders number-6 and -9.

Past and present transportation vehicles and PAA's role in the new age of air travel. Notice that the Convair Camel is the featured "giant airplane" of the soon-to-be-seen future.

Flying down to Rio wasn't just a movie title. Pan American aimed to make it a nonstop reality with giant DC-7s.

As previously alluded to, the aeronautical design world, egged on by the airlines of the United States, was continually dreaming and thus trying to capitalize on the ever-growing capabilities of engine technology as applied to always-bigger airframes. As we know, the range for land planes was limited, necessitating the flying boat. But to get a runway-based airliner to be able to cross the oceans took a large airframe to carry all that fuel.

Always planning for the future, PAA's Juan Trippe was a primary precipitator of the extra-large airplane. At PAA, this seemed a constant and particular compulsion and preoccupation of its president. Shown here are some of the "giant" airplanes he had a hand in with the manufacturers who were developing these designs initially or concurrently for the military services. Some were outlandish and never came to commercial production fruition and would have definitely been too large for the market anyway. Others did make it to market and changed the way international airline flying matured. You can suppose too that overall, it was the airplane's grandeur and not the operability that he was striving to associate with his airline.

Starting in 1942, PAA instigated commercial application design work at Boeing and Douglas for what became the C-97 (Stratocruiser) and C-74 (first DC-7), both entering military service and one into airline work.

Meanwhile, at Convair, the B-36 bomber received a double-deck fuselage to become the XC-99. Working with engineers at Lockheed, its Model 89 was underway. The XC-99 was known as the Camel and the 89 as the Constitution. Both flew and entered military service but not with the airlines. Only one Camel was produced for the air force and two R6O-1s lived lives in the navy, all flying successfully for a few years.

These latter two airplanes would have been the 747s or A380s of their day, carrying 200 passengers enjoying amenities to make the ocean liner *Queen Mary* blush.

Douglas DC-7

If you take the USAAF Douglas C-74 with its heavy lift capability and civilianize it, you get the first iteration of the DC-7. PAA was always desperate to acquire and operate first the biggest and best airframe on the market.

Flying past the Palos Verdes peninsula near Los Angeles, the only XC-99 Camel looks gigantic with its double-deck fuselage. Flying this airplane on a daily basis with all the necessary load factor required would have been a financial and maintenance challenge, to say the least.

A lovely photo of a rare airplane with its slab-sided double-deck fuselage. Although never operational, the airplane still exists in static display form today. (Photo Courtesy Gerald Balzer Collection)

This ad shows PAA's color scheme on the airliner version of the Camel. The upper window line was lowered everywhere but over the wing carry-through box, and the forward lower seating area was now being used for something other than passengers. Nice job too of splitting the difference with the blue stripe between levels.

Figure 2–5. Front and Rear Crankcase Bearings and Center Crankcase Assembly

1. Front Main Bearing 3. Crankshaft Front Journal 5. Rear Crankcase
2. Front Crankcase 4. Crankshaft Rear Journal 6. Rear Main Bearing

The first R-2000s used the same crankcase as the R-1830. However, the additional power from the R-2000 proved to be too much for the crankcase, and cracks occurred. (Photo Courtesy Handbook Overhaul Instructions R-2000-4 Aircraft Engines, May 1957)

The placement and drive of various accessories were also different from those of the R-1830. A 0.50 propeller gear reduction was achieved through the use of bevel gears, and the R-2000 was available with either a single- or two-speed supercharger. The engine had a 6.5:1 compression ratio.

The R-2000 was first run in May 1940 and entered limited production in 1941. The engine initially had a takeoff rating of 1,300 hp at 2,700 rpm and a normal rating of 1,100 hp at 2,550 rpm, both on 87-octane fuel. Power was later increased with 100/130 PN fuel to 1,450 hp at 2,700 rpm for takeoff and 1,200 hp at 2,550 rpm for normal operation. The R-2000 had a 49.1-inch diameter and

weighed around 1,570 pounds. Licensed production of the engine during World War II was undertaken by Buick, but it only built a small number (possibly around 600) of R-2000 engines.

The R-2000 was never installed in a large variety of aircraft as was the R-1830 and R-2800, but the engine's timing was right. The DC-4 design was built for the military as the C-54, and more than 1,100 of the transports were built. The R-2000 served reliably, carrying cargo in C-54s and passengers in DC-4s. Total production of the R-2000 was 12,966 engines.

The R-2000 was selected to power the de Havilland Canada DHC-4 Caribou in the mid-1950s after R-2000 production ended. However, the R-2000 was never used in as many aircraft as other contemporary engines. Developed to produce more power from the basic R-1830 engine, the R-2000 did not possess the same level of reliability as its predecessor. Although it served well in the DC-4 and its derivatives, the R-2000 engine required more delicate operation than the R-1830 and was not as well liked as the smaller, more robust engine.

Pratt & Whitney R-2000 Engines in Airliners				
Manufacturer	A/C Model	Engine Model	HP t/o	Number
Douglas	DC-4	2SD1-G	1,450	4
		2SD3-G	1,450	4
		2SD5-G	1,450	4
		SSD13-G	1,450	4
	C-54, B; R5D-1	R-2000-3	1,350	4
	C-54A, B, C, F	R-2000-7	1,350	4
	C-54G, R5D-2, -5	R-2000-9	1,450	4
	C-54D, E; R5D-3, -4	R-2000-11	1,350	4
	R5D-1, -2, -3, -4	R-2000-4	1,450	4
	Super DC-3S	2SD7-G	1,450	2
SAAB	90 Scandia	2SD13-G	1,450	2

Major Applications of the P&W R-2000 Twin Wasp Engine	
Aviation Traders	ATL-98 Carvair* transport
de Havilland Canada	DHC-4 Caribou transport
Douglas	DC-4 (and derivatives) airliner / transport
SAAB	90 Scandia airliner
* Modified from DC-4 and C-54 aircraft	

Chapter 5

EXPANDING THE ENVELOPE

World War II has been previously noted to have spawned a technological revolution in aviation and rocket science. It served also as a demarcation point in time and perceptions for the 20th century. The airline business crossed the threshold from a world of novelty to one of necessity. PAA led the charge into the future with its "what if" airliner designs as it anticipated a new singular world recently uncovered by the erosive winds of war. Following is a discussion of two very different airplanes, one that saw the light of day and one that did not. More's the pity that the loser never materialized as a viable commercial entity.

Getting There by Air

After 1946

After a request for proposals by the Army Air Forces in 1944 for a fast, *very fast*, reconnaissance airplane with incredibly long-range capabilities, two unlikely contenders answered the call: one from the usual location of Southern California and the other from Long Island, New York. Squaring-up against each other were the Hughes Aircraft Company and Republic Aviation Corporation. The unique approaches each brought to the table could not have been farther apart. For Hughes it was a twin-engine airplane reminiscent of a Lockheed P-38 on steroids, designated XF-11 ("F" for photo). For Republic, it was a development concept for a huge four-engine airplane that was similar to just about nothing else.

Clipper America in all its bare-metal glory. Note the recent removal of the "C" for Commercial in the airplane's registration number. (Photo Courtesy Boeing Co. via Jon Proctor with permission)

This is the second of two prototype Republic XR-12 reconnaissance aircraft. When lengthened, it became the RC-2 Rainbow airliner. Note the streamlined shape of the fuselage. (Photo Courtesy Mike Machat Collection)

That's Impossible!

In the world of 1945, it would have seemed pure science fiction to imagine a piston-powered airplane about the size of a B-29 bomber that could cruise the airways at an astonishing 400 mph and do so nonstop just over 4,000 miles and at 40,000 feet. But this is just what the folks at Republic delivered with the XR-12, later known as the Rainbow. Perhaps no other airframe discussed in this book better exemplified the omnipresent theme: Engine technology makes possible all the advances in air travel and aircraft design.

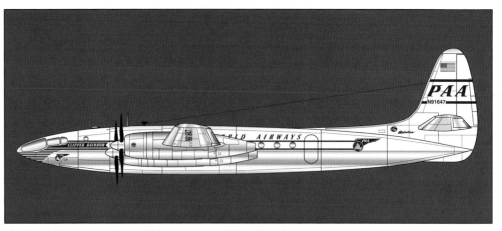

If it had flown . . . it would have looked like this drawing by the talented Tony Landis.

The XR-12 was a perfect aerodynamic shape that took full advantage of the very best of developed piston-engine technology at that time. The engine of choice was P&W's mighty R-4360, developing 3,500 hp. This happy marriage yielded perhaps the most fantastic airplane of the immediate postwar period anywhere in the world.

Juan Trippe seized on this incredible machine and ordered 6 with options for 12 of the airliner version known as the Republic RC-2 Rainbow. (After all the war clouds and their thunderstorms of world combat fighting, the "Rainbow" appellation represented a

This Republic factory model shop Rainbow is in the markings of launch customer Pan American. The long, long tapered nacelles housing the R-4360 engines are quite apparent. (Photo Courtesy Mike Machat Collection)

PAA's cabin configuration shows four-abreast seating, which evidently was a pretty tight fit into the tapered fuselage of the Rainbow. Also interesting to the airliner aficionado is the British-type round sidehead bolsters at the top of each seat. (Photo Courtesy Mike Machat Collection)

The business end of the RC-2 displays a very clean instrument panel and throttle quadrant. The airplane simply wasn't a large aircraft, which was good news in terms of thrust-to-weight ratio and overall speed, of which the Rainbow had plenty. (Photo Courtesy Mike Machat Collection)

On the domestic side of the ledger, co-launch customer American planned on super-fast nonstop transcontinental trips using its Rainbows. Notice the rectangular windows in the passenger cabin in contrast to PAA's round portholes. This square versus round choice also appears with the Stratocruiser.

A cutaway of the P&W R-4360 engine, literally bristling with its 28 cylinders in their helical arrangement for cooling purposes. Four of these on the 110,000-pound Rainbow really gave it some speed. (Photo Courtesy Wikimedia Commons)

This color advertisement from Republic touts its first and only two customers for the RC-2. What an amazing airliner this airplane would have been. (Photo Courtesy Mike Machat Collection)

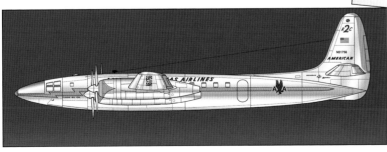

This Tony Landis rendering is of the assumed American paint scheme for the Rainbow. American also used a touch of the air force version cockpit windows, but to tell the truth, the windscreen configuration for the civilian airplane was constantly changing within the design department at Republic.

calming peacefulness following the conflict.) The idea of plying the world's air routes at such speeds and distances was a dream come true as far as land planes were concerned. Finally, flying boats could be properly retired.

Within the sleek tapered fuselage of the Rainbow stretched an additional 5 feet from the military version; 40 passengers and a crew of 7 could be accommodated. The airplane could cruise at an altitude of 40,000 feet with its pressurized cabin registering 10,000 feet for habitability. The Rainbow was approaching jet operating profiles but with props and cylinders. Compare these numbers with the advanced Lockheed Constellation mentioned previously: 300 mph over a range of 3,000 miles but with 60 paying fares. That represents a 33 percent increase in capability over the newest airliner in the skies.

The RC-2 would have entered service a scant two years after the Connie. And when placed head-to-head with the new Douglas DC-4 at its 250-plus mph, well, there just wasn't any comparison.

International flying was by no means the sole applicability for such agility. American Airlines, billing itself as "America's leading airline" wasn't about to let TWA best it going cross-country in the United States and jumped immediately aboard the Republic bandwagon by ordering 20 airplanes. Imagine flying from New York to Los Angeles in *6 hours, nonstop.* Or, as did PAA, AOA traveled from New York to London in about 8 hours. This was a whole new world for the architects of air travel. The DC-6 was on the way from Douglas and it would have been fine as the workhorse of American's routes, but the Rainbow would have set new standards as the premier airplane in this or any airline's fleet. (Republic marketed aggressively to the world's airlines and even had some interest from TWA and KLM.)

The XR-12 had a rather small fuselage cross-section for its military photo mission, which made for a bit of coziness as an airliner, but nevertheless, a club lounge was located over the wing carry-through box and a special women's lounge in the aft of the airplane with sofa seating and vanities. Republic described its "wunderplane" as having fluorescent lighting, inflight movies, and a complete galley. Seats were arranged in pairs, two abreast across the fuselage width.

One modern and unique feature from a ground-handling standpoint was the single-point refueling receptacle located on the lower fuselage near the wing leading edge. This smart innovation was seen later during the Jet Age on airplanes such as the Lockheed Electra, Douglas DC-8, and Boeing 707. Republic was doing everything right with this airplane, or so it had once seemed. Then reality entered the picture.

Each RC-2 would cost the pretty sum of *$1.25 million.* Remember that war-surplus C-54s were going for one-third to one-tenth that amount. Even *new* DC-4s were about one-third the price. Once

the actuarialists and bean counters finished analyzing the unit cost and applied that to the total number of passengers on board, only 40, the per-seat mile cost became astronomical. This airliner would have to run at virtually 100 percent load factor all the time, a known impossibility. That translated in the real world to exorbitant fares paid for the privilege of flying nonstop and in a hurry from East to West coasts. The country, however, was in a typical postwar recession.

By service date entry, inflation was the name of the game for the economy and dollars for such near-frivolity appeared scarce. The Rainbow, had it entered service, was aptly described by author Mike Machat as being Concorde, 30 years early: a premium experience at a premium price.

As the airline environment soured for the viability of the Rainbow, American had to finally withdraw its commitment to the airplane. PAA followed a few months later, although it retained the "Rainbow" name for its premium service from the United States to Europe. By that time, the air force had also killed off all hopes of having a dedicated high-performance photo-reconnaissance airplane, converting existing piston-powered Boeing B-50 and jet-powered B-47 bombers for that role. The Rainbow never flew in commercial service, and not only was the pride of Republic Aviation Corporation grounded but the world lost a truly great airliner in the process.

Luxury and Sound Economics

As all this high drama was taking place in New York, the folks in Seattle were busily designing another cargo plane/airliner around one of their military platforms, this time the B-29 bomber. As with the B-17 to 307 Stratoliner hybrid, designers simply took the wings, tail, engines, and landing gear and grafted them onto a newly designed fuselage. Simple. The result was what Trippe had been prodding them for behind the scenes: a large, long-range airplane that carried many fares. It was the Model 367, and it looked big.

The Model 367 had a two-lobe fuselage: the lower portion was the same diameter as the B-29 fuselage and the upper was larger and quite spacious at close to 11 feet wide. This particular "readymade" version then metamorphosed into a much-improved version just as the B-29 changed into the B-50: bigger tail and the more-powerful R-4360 engine, just as used in the Rainbow and the B-36. Now we were getting somewhere. The commercial airliner version was known as the Model 377, and Boeing chose to stick with its high-altitude moniker system and so called it the Stratocruiser.

The "Strat," as it was known within the industry, was a pure luxury airplane. This design, now going into mass production, was the final distillation of all those giant airliner studies conducted earlier

AOA was the international flag carrier offshoot of American Airlines and was among the first to operate both the Stratocruiser and the 049 Constellation. Juan Trippe took great umbrage against AOA and forced it to be purchased by PAA after the government intervened on behalf of Trippe in a very contentious proceeding. (Photo Courtesy Boeing Co. via Jon Proctor with permission)

The power to lift the giant Boeing rested with, once again, the R-4360. The choice was not only square versus round windows, but also which propeller the customer wanted, either Hamilton Standard seen here (Northwest, PAA) or Curtiss Electric (United, AOA). PAA's choice haunted it by producing propeller failures due to the hollow steel blade of its props wherein the rubber filler shifted and caused imbalance, thus throwing the blade off kilter and causing separation. At least three airplanes were lost due to this anomaly.

This view of a United 377 shows the difference square windows make to the look of the Stratocruiser, one of two decisions for an airline customer when ordering the airplane. (Photo Courtesy Jon Proctor)

Here's why the R-4360 engine was nicknamed the "corncob." (Photo Courtesy Gerald Balzer Collection)

by the airlines, most notably PAA. The 377 could carry 60 passengers in a typical all-first-class layout, up to 100 if you made it an Air Coach. It cruised at 340 mph, sprinted at 375 mph, and could travel 4,200 miles if it had the extra fuel tankage that PAA installed. The altitude maximum registered 32,000 feet when light in weight.

All these numbers were nudging at the forecast envelope of the Rainbow, which had come up short by having to pay the price for economic inefficiency. The list price was also higher than the RC-2, at $1.5 million, but many more passengers made up the difference in the bottom line. Much like today's A380 super jumbo, however, the Strat was a niche airplane and so only 55 were built, thus saturating the small market demand.

Nothing Like It

Reminiscent of the flying boat era just being eclipsed, the new super-landplanes continued to require accommodations for their passengers based upon trip times that were many hours in duration. Much of the inflight experience was carried over from the boats and updated in the new airliners.

With the size of airplane available, all those experiments of the 1930s could now become practical realities. One small example was United's desire to provide a single segregated stateroom for its honeymooning couples. This idea, you may remember, was first introduced on the DC-4E. On the Stratocruiser, this separate compartment was located at the extreme rear of the fuselage (this area was used by PAA for its large galley) and could be booked specifically by a travel agent. United ran its fleet from California to Hawaii, a common routing for newlyweds.

Having taken space from the rear galley layout, United moved its galley amidships and could set it out as a buffet for light fare during the snack service period of the flight. Flying by Strat to the islands was akin to walking the promenade deck of any of the ocean liners of the day. And if that weren't enough of an excursion, there was always downstairs.

Yes, perhaps the most unique feature of the Stratocruiser, used by all airlines operating the type, was the downstairs lounge. Remember that the 377 had an upper and lower fuselage system and thus had

Original paint scheme for the Northwest 377 fleet, bare-metal airliners later gave way to white-crown fuselages at the behest of the Defense Department, which wanted international airliners approaching the country from the oceans to be visually discernable from Soviet metal bombers. (Photo Courtesy Boeing Co. via Jon Proctor with permission)

Later in its career, a white-top Strat is seen getting ready for work at Chicago's Midway airport. Notice the Wisconsin Central DC-3. This airline later became North Central, then Republic Airlines. Just above the DC-3 can be seen a TWA Martin 202 taxiing in to the terminal area. (Photo Courtesy Air Classics Magazine)

This Northwest lounge has a slightly different bar configuration. The bottom floor between the seats is actually the curved underside of the fuselage.

plenty of room not just for baggage, but for something "special." Just aft of the wing was carved out a seating area with a small galley/wet bar at the opposite end. This area could seat up to 14 people in a horseshoe arrangement of seats, often with a table in the center just like a booth at your favorite restaurant. The airline used either a steward or stewardess to serve in this area, so the party was on!

The lounge was reached via a circular staircase next to the main cabin door. The lounge had its own windows and on some models an entry door. (Note how Boeing inverted this idea in the 747, moving the lounge *up the stairs*. Having even more room, it was just as popular in the 1970s as 20-some years before.)

Once all the meal and cocktail service had completed, it was time to get ready for bed. So off to the spacious and well-appointed women's or men's dressing areas and washrooms to put on pajamas and slippers and brush teeth. Again, and as promised in the prior decade, these areas were large enough to suit any size passenger and allowed for all the

Martini anyone? The camera is poised on a shelf next to the steward who is behind the bar mixing cocktails (you can see him in the cross-cabin mirror). Notice the midcentury pink of the wall covering. (Photo Courtesy Boeing Co. via Jon Proctor with permission)

Just aft of the main passenger door lay the circular staircase that connected the main deck seating area with the spectacular downstairs lounge. In this staged photo aboard a PAA airplane, the young woman is just emerging from her climb to the main floor. Above the heads of the passengers are the stowed fold-down berths. (Photo Courtesy Boeing Co. via Jon Proctor with permission)

And sleep is just what is in mind with our passenger, as she relaxes with some reading as the "smooth highway in the sky" lulls her into blessed somnolence.

civilized behavior you could hope for while traveling. No more stopping at waypoints or islands to spend the night in a hotel. Your accommodations traveled with you in the new air age.

Sleeping arrangements were of paramount concern on overnight flights, and, as is still the case today in First Class and Business Class long-haul flying, the bed was an integral part of the traveling experience. At PAA, the Stratocruiser fleet came equipped with pull-down berths located above the daytime seating, once again à la Pullman style. Once lowered, curtains were attached to create private compartments. Below, the seats were reclined and curtains erected around them as well. Unlike Douglas, Boeing chose not to

Remember the incredible thought of a hotel-style lavatory on the DST in 1936? Our Stratocruiser bathroom is reminiscent of something found on a steamship. Now brush your teeth and let's go to sleep, Junior.

With the Pullman berths lowered, the seats reclined and were made up with linens, curtains ready to be drawn for privacy. The kids received some hot chocolate or milk prior to turning in onboard this PAA Stratocruiser, bound for the far-flung corners of the Earth. Mom and Dad can finally escape downstairs for a nightcap and some cards.

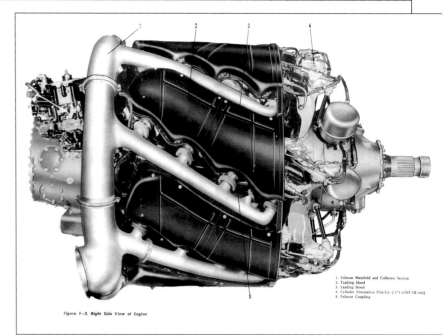

Figure 1—2. Right Side View of Engine

1. Exhaust Manifold and Collector Section
2. Trailing Hood
3. Leading Hood
4. Cylinder Detonation Pick-Up
5. Exhaust Coupling

The R-4360 incorporated extensive hooded baffles to direct cooling air through the cylinder's cooling fins. The baffles combined with the engine's helical configuration to ensure proper cooling of all cylinders. (Photo Courtesy Aircraft Engine Historical Society)

incorporate small upper-berth windows for sleeping passengers, so you were cozily cocooned within the bed system. The stewardess crew members certainly had their work cut out for them, making all these beds. You can imagine the passengers milling about while seats were disrupted and beds lowered. No matter, this was still luxury aloft.

Pearce

Hardware on the Wing

Pratt & Whitney R-4360 Wasp Major

In the late 1930s, P&W was involved with the development of a series of experimental liquid-cooled engines. The Army Air Corps and US Navy sponsored the engines, which were the 1,800-hp H-2600, the 2,700-hp H-3130, and the 3,000-hp H-3730. All of these engines used 24 cylinders in an H configuration and employed sleeve valves. George Mead was in charge of the liquid-cooled engine projects, which showed some promise but required much more development.

Suffering from poor health, Mead resigned from P&W on 27 June 1939. Leonard (Luke) Hobbs succeeded Mead as P&W's chief engineer. Don Brown, president of P&W, unexpectedly passed away on 29 January 1940. In seven months, the company had lost two of

its founders and was understandably overwhelmed with completing the multitude of tasks required to develop the R-2800 and R-2000 engines, fostering various experimental programs, and producing its proven engines. P&W needed direction.

Since 1930, Frederick Rentschler had been heading P&W's parent company, United Aircraft Corporation. With the difficulties facing P&W, Rentschler returned in 1940 to lead the company through the troubled times. Hobbs recommended to Rentschler that the liquid-cooled engines be canceled, as P&W's experience was with air-cooled engines. Hobbs had drawn up some proposals of how to configure a 3,000-hp air-cooled engine, which included three rows of 9 cylinders (27 cylinders total), five rows of 7 cylinders (35 cylinders total), six rows of 5 cylinders (30 cylinders total), and six rows of 6 cylinders (36 cylinders total). However, the primary focus was on an engine with four rows of 7 cylinders (28 cylinders total).

P&W had been contracted to build the liquid-cooled engines and needed permission to cancel development of those projects in order to focus on Hobbs's 3,000-hp air-cooled engine. Rentschler met with General Henry (Hap) Arnold, chief of the Army Air Corps, to discuss the air-cooled engine. Both men knew that war was coming soon. Rentschler explained the engine situation and said that a 3,000-hp liquid-cooled engine had years of development ahead of it

Perhaps the most romantic trip for midcentury travelers in 1950s America was a vacation to the tropical paradise known as Hawaii. PAA had much to do with this, as discussed in chapter 3, and now it was the turn of both PAA and United Air Lines to ply the route but using its brand-new cruise liner of the skies, the Stratocruiser. Service emanated from both San Francisco and Los Angeles and took as long as 12 hours flying against the winter headwinds. Not since those flying boat days had there been such luxury aloft for passengers.

The Stratocruiser was obviously the preeminent airplane of its day. It broke the barrier of long-distance land plane incapabilities and allowed, for a brief decade, the ultimate passenger experience in getting there by air.

Of note in the United airplane was the honeymoon suite at the rear of the airplane and the long galley in the center that was made into a serving buffet for a second meal during the long flight to the islands, passengers helping themselves to the food arrangements thereon.

Inflight near Seattle, Washington. (Photo Courtesy Boeing Co. via Jon Proctor with permission)

Front office of the 377: steady hands guiding the sky ship safely to its destination. (Photo Courtesy Alan van Winkler via Jon Proctor)

Honolulu International Airport

We certainly must be in Hawaii because the tower offers a hearty "Aloha."

NOW! DIRECT SERVICE VIA LOS ANGELES TO HAWAII

United Air Lines Mainliners

United's new direct route via Los Angeles is in addition to our San Francisco-Honolulu route. Now you can go by way of one city and return via the other at no extra cost! Leave the East by noon, be in Hawaii for breakfast! Or leave in late evening, be there for dinner! Enjoy the luxuries of the world's finest Stratocruiser*. For reservations, low-cost Hawaiian vacations, or a sea-air holiday (one way by Mainliner, the other by Matson or American President ship), call or write United Air Lines or see an Authorized Travel Agent.

*Mainliner Stratocruiser from California, DC-6 Mainliner 300s from the East and Midwest. Stratocruiser berths or staterooms at slight extra cost.

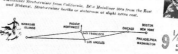

Later in United's exercising of its route rights to Hawaii, it chose to use its new DC-7s to cross the Pacific; faster yes, but not nearly the same as flying on a Boeing. Until introduction of the 747 and DC-10, widebody luxury to the islands was noticeably absent on the route. (Photo Courtesy Jon Proctor)

Standard operating procedure was to be greeted by ukuleles, hula dancers, and flowered leis once stepping foot on Hawaiian soil. Paradise at last!

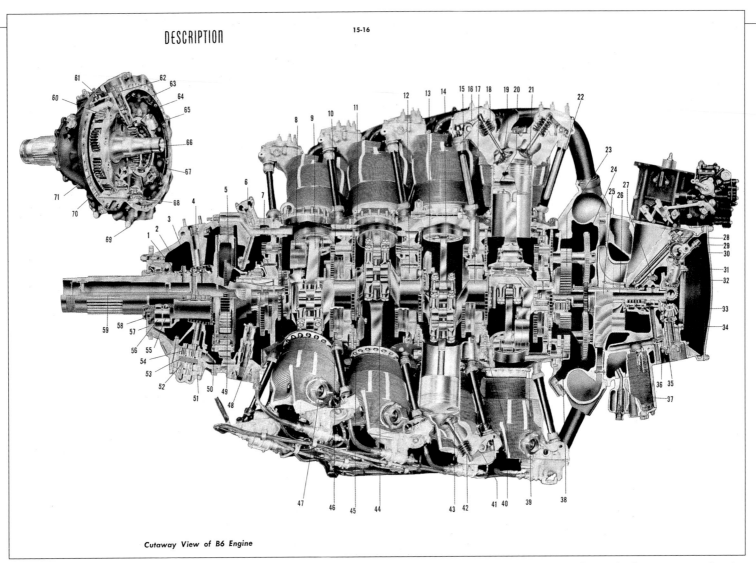

Cutaway View of B6 Engine

1) Oil transfer bearing ring carrier, 2) oil transfer bearing, 3) oil transfer bearing rings, 4) governor drive shaft gear, 5) reduction drive gear outer coupling, 6) spark advance control valve, 7) crankcase oil pressure line, 8) crankshaft front counterweight, 9) D-row master rod assembly, 10) cam small drive gear, 11) cam large drive gear, 12) crankshaft center main bearing, 13) crankshaft oil slinger, 14) B-row link rod, 15) inlet valve pushrod, 16) inlet valve rocker, 17) crankshaft, 18) inlet valve springs, 19) inlet valve, 20) piston pin, 21) piston and rings, 22) pushrod cover, 23) impeller drive damper, 24) impeller intermediate drive, 25) impeller and shaft assembly, 26) accessory drive shaft, 27) impeller shaft rear rings breather, 28) fuel feed valve, 29) tach drive gear, 30) fuel pump drive gear, 31) fuel pump intermediate drive, 32) rear oil distributor ring, 33) starter drive gear, 34) generator or accessory drive gear, 35) rear accessory drive gear, 36) rear accessory drive oil pressure reducing valve, 37) pressure oil strainer, 38) collector case oil pump, 39) crankcase scavenge oil line, 40) crankcase scavenger oil pump, 41) exhaust valve rocker, 42) exhaust valve springs, 43) exhaust valve, 44) C-row master rod, 45) cam drive gear, 46) cam, 47) exhaust port flanged coupling, 48) magneto intermediate drive gear, 49) propeller shaft reduction drive gear, 50) front accessory drive gear, 51) torquemeter pump, 52) front power section scavenge pump, 53) spacer section, 54) front section scavenge pump, 55) propeller shaft ball bearing, 56) propeller shaft roller bearing, 57) thrust cover, 58) propeller shaft thrust nut, 59) propeller oil feed tube, 60) propeller shaft reduction drive fixed gear, 61) torquemeter oil pressure transmitter, 62) magneto drive shaft, 63) spark advance oil feed tube, 64) magneto drive fixed gear, 65) spark advance cylinder, 66) propeller shaft, 67) magneto drive gear, 68) torquemeter master piston, 69) torquemeter oil pressure relief valve, 70) propeller shaft reduction pinion support, and 71) propeller shaft reduction pinion. (Photo Courtesy Overhaul Manual Wasp Major Models B6 and CB2, December 1957)

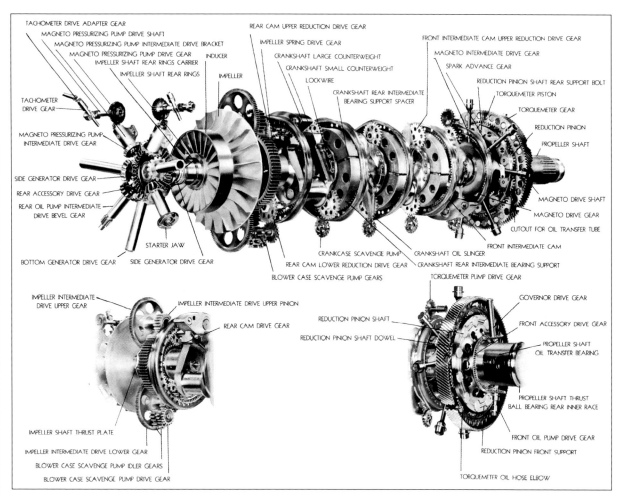

The complex internals of the R-4360 illustrate the intensive maintenance needs of the engine, which was one of the reasons it was never popular with the airlines. Note the ring gear rear accessory drive that was unique to the R-4360. (Photo Courtesy Maintenance Manual Wasp Major (R-4360) TSB3G and VSB11G Engines, December 1957)

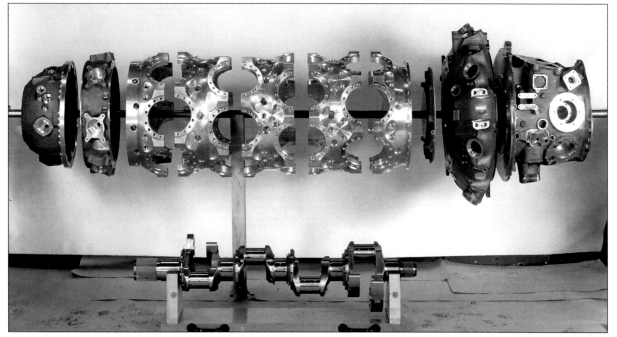

The five forged aluminum sections that make up the R-4360's crankcase are flanked by the cast-magnesium gear reduction and front accessory case (far left) and the supercharger and accessory sections (far right). The three intermediate crankcase sections (center) are identical. The engine's large, one-piece crankshaft is at the bottom. (Photo Courtesy Aircraft Engine Historical Society)

and would probably miss the war. However, P&W's expertise with air-cooled engines would lessen development time of a 3,000-hp engine of that type. If the liquid-cooled projects were canceled and their resources allocated to the air-cooled engine, it would be ready sooner and might see action.

Rentschler stressed that air-cooled engines were simpler, less expensive, more reliable, and more durable than liquid-cooled engines, and he stated that P&W was willing to cover the costs of canceling the liquid-cooled engines. Representing the Army Air Corps, General Arnold approved Rentschler's plan. A short time later, the navy also agreed; air-cooled engines were always the navy's preference. On 11 November 1940, design work on the 3,000-hp X-Wasp engine was initiated. The X stood for "experimental" and was not the engine's configuration.

The X-Wasp used 28 cylinders that were based on cylinders from the R-2800 engine. The 5.75-inch bore and 6.0-inch stroke gave the X-Wasp a displacement of 4,362 ci, and the engine was untimely designated the R-4360 Wasp Major. Hobbs set an aggressive schedule for the first proof-of-concept engine, and it was built using many components from other engines. The prototype engine, called the X-103, used the front cylinders and supercharger from the R-2800 B-series, the nose case and gear reduction from the H-3130, and connecting rods from the R-2180. The X-Wasp had four cam rings: one positioned at the front of the engine and one between each row of cylinders. Each cam ring served the row of cylinders immediately behind it.

The cylinders on the same row were spaced the normal amount for a 7-cylinder radial, which is $51\tfrac{3}{7}$ degrees. Dividing $51\tfrac{3}{7}$ by four (the number of rows) gave the $12\tfrac{6}{7}$-degree stagger that each row had from the row preceding it. This helical cylinder spacing (and extensive baffling) allowed air to flow between the cylinders of each row and to the cylinders in the next row so that even the rear row of cylinders received adequate airflow for cooling. The four helical rows of seven cylinders gave the R-4360 some resemblance to a corncob, which is how the engine got its nickname: the corncob radial.

The cylinder rows were labeled starting with the A row at the rear of the engine. Moving forward along the engine, the next rows of cylinders were the B row and C row. The D row were the cylinders at the front of the engine. The cylinders were numbered clockwise when viewed from the rear, starting with their row designation and followed by the cylinder number. Cylinder A1 was the top cylinder on the rearmost row. The next clockwise cylinder was B1, followed by C1, and finally D1, on the front row. The next cylinder was A2, back on the rear row, and the numbering continued around the engine. The master connecting rods were located in cylinders A7, B4, C4, and D1.

The R-4360's crankcase consisted of five forged aluminum sections: a front crankcase section, three intermediate crankcase sections, and a rear crankcase section. All crankcase sections were split vertically though the cylinders. The two-piece gear reduction housing and the two-piece supercharger and accessory housing were made of cast magnesium. The engine used a single-piece, forged steel crankshaft with four throws, each spaced at $192\tfrac{6}{7}$ degrees (180 degrees plus the $12\tfrac{6}{7}$-degree stagger of the cylinder rows). The crankshaft main bearing carriers that supported the crankshaft in the three intermediate crankcase sections were made of magnesium. The crankshaft used silver-lead-indium bearings for its main journals and crankpins. The two-piece master connecting rods were a similar design to the master rods used on the R-1830 engine. For even fuel distribution to the engine's 28 cylinders, a slinger ring was employed to deliver fuel just before the supercharger's impeller.

The prototype R-4360 was first run on 28 April 1941. The early engines were known as the A-series and were developmental engines. The A-series produced 2,800 hp at 2,600 rpm for takeoff and had a normal rating of 2,400 hp at 2,400 rpm on 100-octane fuel. The engine had a diameter of 51.75 inches and weighed around 3,200 pounds. A-series engines accumulated 500 hours of test runs in 1941.

In March 1942, the navy ordered an R-4360 engine. Limited production of additional developmental engines began in April 1942. These preproduction R-4360s incorporated several changes from

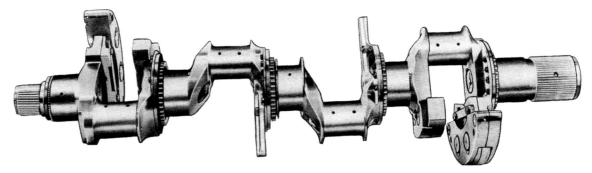

The crankpins on the R-4360's crankshaft were staggered at $192\tfrac{6}{7}$-degree intervals to accommodate the stagger of the cylinder rows. The geared sections of the crankshaft drove the cam ring drives. (Photo Courtesy Maintenance Manual Wasp Major (R-4360) TSB3G and VSB11G Engines, *December 1957)*

the previous engines and were the start of the B-series. The R-4360 B-series had a takeoff rating of 3,000 hp at 2,700 rpm and a normal rating of 2,500 hp at 2,550 rpm using 100/130 PN fuel. The engine had a compression ratio of 7.0:1, a diameter of 52.5 inches, and weighed approximately 3,500 pounds.

Production B-series engines differed from the earlier prototype engines in several ways. No longer were components borrowed from other engines; production R-4360s used parts made especially for it. The R-4360's cylinder had the intake valve positioned in the front of the cylinder and the sodium-cooled exhaust valve in the rear. As installed, the cylinders were rotated 30 degrees counterclockwise on the crankcase. This configuration created seven densely packed helices of cylinders but also increased space between cylinders on the same row. The extra space created additional cooling airflow that was directed through baffles toward the cylinders. The airflow through the baffles was at 30 degrees to match the cylinders. Initially, open baffles were used, but this was switched to hooded baffles that enclosed the cylinders and aided cooling.

The intake and exhaust manifolds spiraled along the engine with the cylinders. The intake manifold was positioned above the cylinders, which allowed more air to flow between the cylinders. The intake manifold fed each cylinder via its downdraft intake port. The cylinders consisted of a forged aluminum cooling muff shrunk

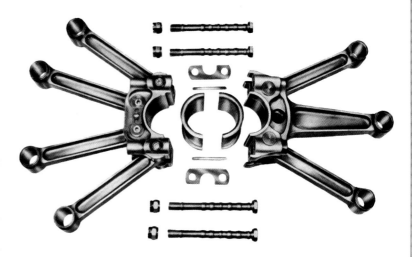

The master and link connecting rod arrangement used in the R-4360 was essentially the same design used in the R-1830. The two-piece master rod was not as strong as a one-piece master rod, but its use simplified the crankshaft's design, which was quite complex for the R-4360. (Photo Courtesy **Maintenance Manual Wasp Major (R-4360) TSB3G and VSB11G Engines,** *December 1957)*

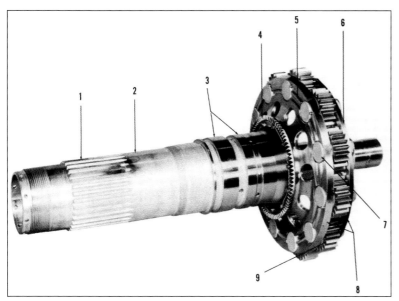

The R-4360 used a planetary gear reduction for the propeller. **1) SAE# 60A spline, 2) propeller shaft, 3) oil transfer bearing support, 4) front accessory drive gear, 5) pinion shaft front support, 6) pinion shaft rear support, 7) pinion shaft, 8) pinion cage, and 9) pinion.** *(Photo Courtesy* **Maintenance Manual Wasp Major (R-4360) TSB3G and VSB11G Engines,** *December 1957)*

onto a forged steel barrel that was screwed and shrunk into a forged aluminum head. All of the R-4360's cylinders were interchangeable and reversible for pusher installations, but the cam rings needed to be changed for pusher installations.

The R-4360 B-series used five identical cam rings to actuate the engine's 56 valves. The cam rings had two tracks, and each track had three lobes. The cam ring at the front of the engine operated only the intake valves of the front row, and the rear cam ring operated only the exhaust valves of the rear row. The remaining three cam rings, which were positioned between the cylinder rows, operated the exhaust valves of the cylinders immediately in front of the cam ring and the intake valves of the cylinders immediately behind the cam ring. The cam rings were driven from their inner edge (internal teeth) and rotated in the opposite direction of the crankshaft at 1/6 crankshaft speed. Split gears attached to the crankshaft drove the middle three cam rings via intermediate gears.

The most common propeller gear reduction ratio used on the R-4360 was 0.375, but 0.381 and 0.425 were also widely used. Gear reduction was achieved through planetary gears. Accessories mounted around the accessory section at the rear of the engine were driven by a single ring gear. This configuration shortened the engine and made accessing, replacing, and switching the accessories easier. It also left the back of the engine free to power a remote supercharger or an auxiliary cooling fan for pusher installations. Most

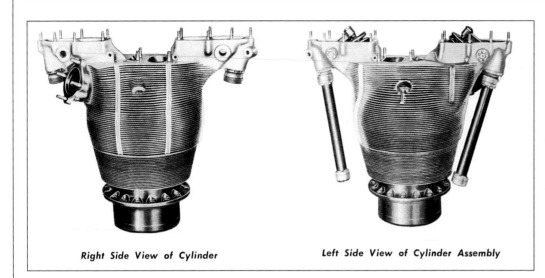

The front and rear valve locations on the R-4360 cylinder were unique and a result of the engine's four-row design. Mounted to the engine, the front valve was the intake, which was fed through the port at the top of the cylinder. The rear valve was the exhaust, which led to the port on the back of the cylinder. (*Photo Courtesy Maintenance Manual Wasp Major (R-4360) TSB3G and VSB11G Engines, December 1957*)

Right Side View of Cylinder   *Left Side View of Cylinder Assembly*

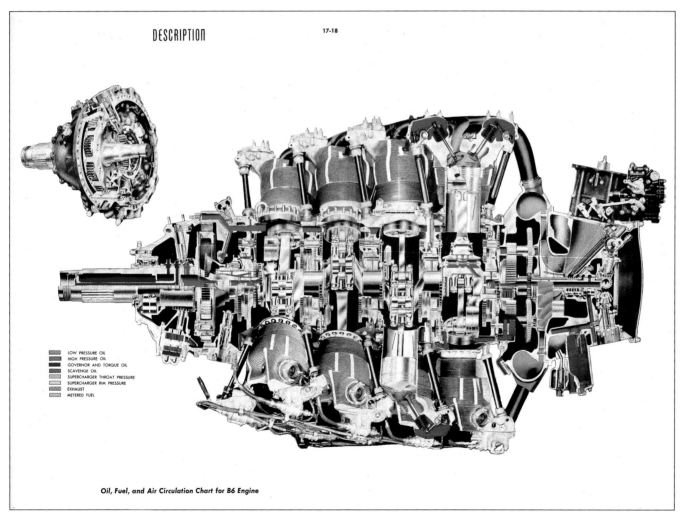

DESCRIPTION   17-18

LOW PRESSURE OIL
HIGH PRESSURE OIL
GOVERNOR AND TORQUE OIL
SCAVENGE OIL
SUPERCHARGER THROAT PRESSURE
SUPERCHARGER RIM PRESSURE
EXHAUST
METERED FUEL

Oil, Fuel, and Air Circulation Chart for B6 Engine

In one second at maximum power and 2,800 rpm, the R-4360 produced more than 3,500 hp by consuming 91.5 cubic feet of induction air mixed with 13 fluid ounces of fuel and 4.6 fluid ounces of anti-detonation injection fluid. Almost 1.1 gallons of oil were pumped through the engine, and the propeller rotated 17.5 times. (*Photo Courtesy Overhaul Manual Wasp Major Models B6 and CB2, December 1957*)

R-4360 engines used a variable-speed single-stage supercharger. Some experimental models did have two-stage supercharging created by adding an auxiliary supercharger. The aircraft for which these engines were intended never entered production, and production engines meant for high-altitude operations used the single-stage supercharger supplemented with turbocharging.

In January 1942, a R-4360 B-series engine was shipped to Vultee's factory in Downy, California, for installation in a Vultee V-85 test aircraft. The V-85 was a modified A-31 Vengeance (V-72) dive bomber that had been specially built for the R-4360. The V-85 had extended fixed gear to allow clearance for a larger propeller, oil coolers where the landing gear fairings used to be, and an extended nose to house the massive R-4360.

The V-85 and R-4360 made their first flight on 25 May 1942, and then the aircraft and engine were shipped back to

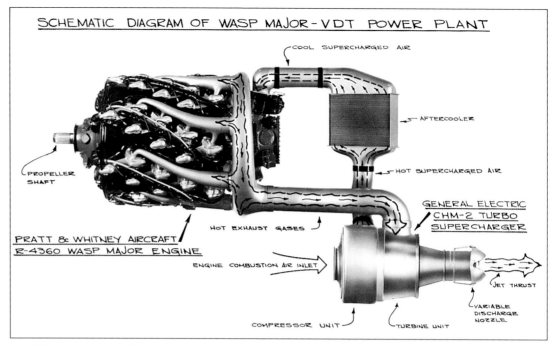

At 4,800 hp, the R-4360 with a Variable Discharge Turbine (VDT) was P&W's next step to outpacing Wright by producing a more powerful and economical piston engine. However, P&W determined that its resources were better spent developing jet engines, and only a few experimental VDT engines were built. (Photo Courtesy Aircraft Engine Historical Society)

P&W's facility in East Hartford, New Jersey. The V-85 flew again on 24 July 1942. After accumulating 5.5 hours of flight time over 14 flights, the aircraft was destroyed while attempting an emergency landing on 15 September 1942. The engine had quit in flight due to a fuel pump failure, not because of a problem with the engine's internals. The pilots were not seriously injured in the crash, but it did delay further flight tests of the R-4360 engine. By the end of 1942, R-4360 test engines had accumulated 3,500 hours of operation.

In mid-1943, the first R-4360 B-series engine was shipped to the military. It was a preproduction model intended for type testing and was received by the Army Engineering Lab in Philadelphia, Pennsylvania. Other preproduction engines were shipped to various aircraft manufacturers starting in September 1943. Eleven R-4360 preproduction engines had been shipped in 1943, and the total run time of test engines was at 7,200 hours. In 1944, 27 preproduction engines were shipped, and standard R-4360 production began by the end of the year. Test engines had accumulated a total of 12,000 hours.

Deliveries of production R-4360 engines began in January 1945, and the engine completed its 150-hour type test on 6 February 1945. In mid-1945, more power from the R-4360 was desired. Certain B-series models had the compression ratio reduced to 6.7:1, which allowed for increased boost. Using 115/145 PN fuel, the engines had

a takeoff rating of 3,500 hp at 2,700 rpm with anti-detonation injection (ADI) and 3,250 hp without. The engine's normal power rating was 2,650 hp at 2,550 rpm. The more powerful B-series were 54 inches in diameter and weighed 3,530 pounds. P&W delivered 114 engines in 1945, and test engines had accumulated 17,000 hours by the end of the year.

To push power beyond 3,500 hp, a complete redesign of the R-4360 was needed, which led to the C-series engines. Design work on the C-series started in November 1945 and carried into 1946. In general, components were strengthened to withstand the stresses of creating additional power. Each cam ring on the C-series had four-lobe tracks and operated at 1/8 crankshaft speed. The cam rings were driven from their outside edge (external teeth) and rotated in the same direction as the crankshaft. The front and rear cam rings only had one track. Cylinders were revised with additional cooling fins, and direct fuel injection was available. Some later B-series engines were retrofitted with C-series power sections.

Using 115/145 PN fuel, C-series engines had a takeoff rating of 3,800 hp at 2,800 rpm with ADI and 3,500 hp without. The engine had a normal rating of 2,800 hp at 2,600 rpm. The C-series weighed approximately 3,800 pounds and was typically 55 inches in diameter. By the end of 1946, test engines had accumulated 24,000 hours,

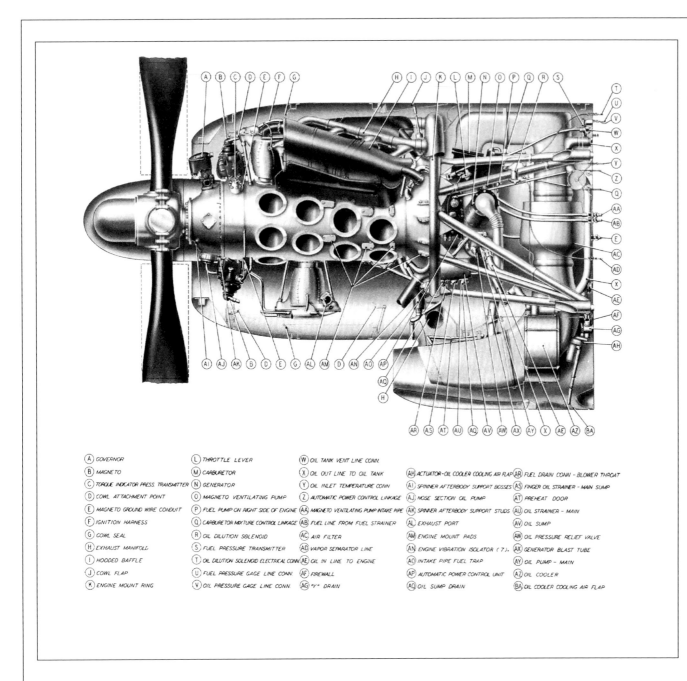

With a combined weight of around 4,300 pounds, mounting an R-4360 engine and its propeller was a serious feat of engineering, especially considering all the vibrations associated with engine operation and the load factors associated with flying. (Photo Courtesy Installation Handbook, March 1948)

(A) GOVERNOR
(B) MAGNETO
(C) TORQUE INDICATOR PRESS. TRANSMITTER
(D) COWL ATTACHMENT POINT
(E) MAGNETO GROUND WIRE CONDUIT
(F) IGNITION HARNESS
(G) COWL SEAL
(H) EXHAUST MANIFOLD
(I) HOODED BAFFLE
(J) COWL FLAP
(K) ENGINE MOUNT RING

(L) THROTTLE LEVER
(M) CARBURETOR
(N) GENERATOR
(O) MAGNETO VENTILATING PUMP
(P) FUEL PUMP ON RIGHT SIDE OF ENGINE
(Q) CARBURETOR MIXTURE CONTROL LINKAGE
(R) OIL DILUTION SOLENOID
(S) FUEL PRESSURE TRANSMITTER
(T) OIL DILUTION SOLENOID ELECTRICAL CONN.
(U) FUEL PRESSURE GAGE LINE CONN.
(V) OIL PRESSURE GAGE LINE CONN.

(W) OIL TANK VENT LINE CONN.
(X) OIL OUT LINE TO OIL TANK
(Y) OIL INLET TEMPERATURE CONN.
(Z) AUTOMATIC POWER CONTROL LINKAGE
(AA) MAGNETO VENTILATING PUMP INTAKE PIPE
(AB) FUEL LINE FROM FUEL STRAINER
(AC) AIR FILTER
(AD) VAPOR SEPARATOR LINE
(AE) OIL IN LINE TO ENGINE
(AF) FIREWALL
(AG) "Y" DRAIN

(AH) ACTUATOR-OIL COOLER COOLING AIR FLAP
(AI) SPINNER AFTERBODY SUPPORT BOSSES
(AJ) NOSE SECTION OIL PUMP
(AK) SPINNER AFTERBODY SUPPORT STUDS
(AL) EXHAUST PORT
(AM) ENGINE MOUNT PADS
(AN) ENGINE VIBRATION ISOLATOR (7).
(AO) INTAKE PIPE FUEL TRAP
(AP) AUTOMATIC POWER CONTROL UNIT
(AQ) OIL SUMP DRAIN

(AR) FUEL DRAIN CONN. - BLOWER THROAT
(AS) FINGER OIL STRAINER - MAIN SUMP
(AT) PREHEAT DOOR
(AU) OIL STRAINER - MAIN
(AV) OIL SUMP
(AW) OIL PRESSURE RELIEF VALVE
(AX) GENERATOR BLAST TUBE
(AY) OIL PUMP - MAIN
(AZ) OIL COOLER
(BA) OIL COOLER COOLING AIR FLAP

and 387 engines were delivered during the year. For 1947, R-4360 production totaled 985 engines, 195 of which were for commercial aviation. Test engines had accumulated a total of 31,600 hours.

P&W investigated several ways to continue to improve the R-4360's performance, including turbo-compounding. However, P&W decided to pursue another path, and no turbo-compound R-4360 engines were built. At takeoff power, some 500 hp was used to power the supercharger on the R-4360 C-series. In 1947, P&W began work to eliminate the supercharger and outfit the engine with what it referred to as a variable discharge turbine (VDT). The VDT was composed of one or more turbochargers in which the flow of exhaust exiting the turbochargers was throttled by clamshell doors. As the exiting flow was restricted, the flow of incoming air decreased. The reduction in airflow was sensed by a master control unit, which decreased the direct-injection fuel flow to the cylinders. The entire engine's output was governed by the VDT's exhaust.

The R-4360 VDT produced 4,300 hp at 2,800 rpm for takeoff and had a normal rating of 3,150 hp at 2,600 rpm. In addition, 300 pounds of thrust was provided by the VDT's exhaust jet. R-4360 VDT engines were intended for the Boeing B-50C / B-54

Superfortress, Convair B-36C Peacemaker, and Republic F-12C / R-12A Rainbow. The VDT engine was first flown on 26 May 1948, when it was installed in the number-3 nacelle of a B-50. This engine had a single two-stage CHM-2 turbocharger acting as the VDT. Flight testing went well, but as events played out, this was the only VDT engine flown, as all of the VDT-powered aircraft projects were canceled.

A further development of the VDT had twin two-stage CHM-2 turbochargers operating in parallel, feeding a single-stage CH-8 turbocharger. This engine demonstrated the ability to maintain 3,000 hp up to 50,000 feet. Advanced VDT engines would have had a takeoff rating of 4,800 hp at 2,800 rpm with ADI and 300 pounds of thrust. This power would have been achievable up to 25,000 feet. Without ADI, the takeoff rating was 4,500 hp with 245 pounds of thrust, and this power could be maintained up to 27,000 feet. The engine's normal rating was 3,500 hp at 2,600 rpm with 300 pounds of thrust at 30,000 feet. It was P&W's goal to obtain a specific fuel consumption of 0.375 pounds of fuel per horsepower per hour with the VDT. Although this was slightly lower than the Wright R-3350 Turbo-Compound, the value was never achieved because VDT development was stopped. Around 24 VDT engines were produced, from 1947 through 1950. The VDT engines were 55 inches in diameter and weighed 3,900 pounds.

R-4360 production for 1948 amounted to 1,220 engines, of which 215 were for commercial use. This represented the high point for commercial sales of the R-4360 engine. Test engines had accumulated more than 40,000 hours since the beginning of R-4360 development. In 1949, 1,576 engines were produced, and only 7 were for commercial sales.

With its R-2800 experience, Ford won a postwar contract to build R-4360 engines along with P&W. As fate would have it, Ford acquired the Dodge Chicago plant that was used for Wright R-3350 production during the war. Ford produced approximately 3,079 R-4360 engines from 1951 through 1954. R-4360 production ended in July 1955, after 18,679 engines were built.

The Boeing 377 Stratocruiser was the only US airliner that was powered by R-4360 engines. The 377's main mechanical issue was not the engine; it was the propeller. Steel propellers with hollow blades were originally fitted to some 377 aircraft. On occasion, a blade broke, leaving an incredibly unbalanced three-blade unit still spinning and attached to the engine. The vibrations ripped the engine from the wing, causing the aircraft to violently yaw. If an experienced flight crew had sufficient altitude, the aircraft could be recovered. Without

Pratt & Whitney R-4360 Engines in Airliners				
Manufacturer	A/C Model	Engine Model	HP t/o	Number
Boeing	377	TSB3-G	3,500*	4
		B6**	3,500*	4
		CB2**	3,500*	4
* With ADI				
** Conversions from TSB3-G engines				

Major Applications of the P&W R-4360 Wasp Major Engine	
Boeing	Model 377 Stratocruiser airliner
	B-50 Superfortress heavy bomber
	C-97 Stratofreighter transport
	KC-97 Stratofreighter in-flight refueling aircraft
Convair	B-36 Peacemaker strategic bomber
Douglas	C-74 Globemaster transport
	C-124 Globemaster II transport
Fairchild	C-119 Flying Boxcar transport
Goodyear	F2G Corsair fighter
Martin	AM Mauler attack aircraft
	P4M Mercator maritime recon aircraft
Northrop	XB-35 flying wing heavy bomber
Sud-Est	SE.2010 Armagnac airliner

an experienced flight crew or sufficient altitude, the aircraft and all on board were lost. After a few aircraft crashed because of this issue, propellers with solid blades were installed on all 377s.

Although it performed well and was noted for its smooth operation, the P&W R-4360 Wasp Major was never very popular with the airlines. R-4360 engines for commercial applications produced 3,500 hp for takeoff, which was about the same power as the Wright R-3350 Turbo-Compound. On a cylinder-by-cylinder basis, performing general maintenance on an R-4360 was easier than performing general maintenance on an R-3350. But, the fact that the R-4360 had 10 additional cylinders with 20 additional spark plugs and valves in two additional rows negated any gains made by the engine's simpler construction over that of the R-3350. The R-4360 had a time between overhauls that only stretched to around 1,400 hours, and the engine had high oil consumption rates. The R-4360 was simply too expensive and required too much maintenance for airline operations.

Chapter 6

SOME SERIOUS AIR TRANSPORTS

Lockheed Model 749A Constellation N6014C Star of Delaware shows its beautiful postwar lines to the camera. Earlier TWA Connie names were prefixed with Sky Chief until 1950. (Photo Courtesy Jon Proctor)

After World War II came to an end, a lot of industrial might had to be scaled back and/or reapplied to peacetime work. Aviation manufacturers led the pack in terms of looking for something to do with all the personnel and applied science they had honed during the conflict. All this raw energy was redirected at either the futuristic weaponry of the emerging Cold War, including the new space and missile disciplines, or new commercial airliners and civilian aircraft, including helicopters and general aviation airframes. This explains why there was so much advancement from 1945 to 1950 in the airline field of endeavor. These were breakthrough years that set the tone for the next decade and the jet airliners that followed.

Getting There by Air

The 1950s, Let the Games Begin

Southern California was the definitive aerospace capital of the postwar world. And as far as commercial airliner transports were concerned, it was the epicenter of all the moving and shaking that commenced as our arms were laid down. Douglas was the preeminent leader in the field thanks to its DC-3 and DC-4, but upstart

The DC-6 prototype on display for the camera, complete with models, at the Santa Monica factory.

A standard or "straight" Braniff DC-6 flys through a Texas sky.

Delivery day! A joint ceremony for the first DC-6s to both American and United took place in November 1947. Despite the leaden rainy skies, a healthy crowd showed up to launch the postwar era of commercial aviation. (Photo Courtesy Jon Proctor)

Here is the constant section main cabin seating. It was very refined with nice wide seats and a wide aisle. (Photo Courtesy Museum of Flying Collection)

Families were welcomed aboard airliners, typically via the air-line promotion of businessmen taking them along at half-price fares. This is a United Air Lines DC-6 scene.

Lockheed, 20 miles away in Burbank, was suddenly the great usurper of the title. This would never do as far as Santa Monica was concerned, and so one of the great industrial rivalries of all time began. There was jousting and fencing for years into the following decades, even being resurrected in the widebody jetliner era of the 1970s (DC-10 vs. L-1011). The suppliers of airliners to the world were on the field and ready to go at it with one another. Welcome to the opening skirmish between the DC-4/-6 and the Constellations. Competition proved to indeed be a healthy exercise for the airliner trade.

You've already learned about the initial round of one-upmanship: the slow and unpressurized Douglas airplane versus the pressurized and speedy Lockheed triple-tail wonder. In the waning days of the war, Douglas had designed its own pressurized, larger, and faster transport airplane for the air force known as the XC-112. The magic of this airplane was the use of the Pratt R-2800 Double Wasp engine, which had proven itself during the rigorous days of conflict. Once the war concluded, the airplane became known as the DC-6 and was the counter to what Lockheed already had flying in airline service in 1946.

The DC-6 was a more developed version of the DC-4, having a fuselage stretch aft of the wing of 81 inches, beefier landing gear, double-slotted flaps, and of course, a pressurized cabin and cockpit. It entered service with United and American following its certification in June 1947. The DC-6 could carry 48 to 52 First Class passengers, range up to 3,300 miles nonstop, and cruise at a speed of 330 mph as high as 29,000 feet. The Douglas concept for a passenger lounge placed a horseshoe-shaped banquette of seats at the rear of the cabin in the tapered rear portion of the fuselage. All the numbers for this airplane matched those of Lockheed's Constellation, giving airlines a true choice of fleet types and manufacturers.

These two postwar airliners could in fact fly nonstop transcontinental service, the long sought-after goal. But due to the length of time at those airspeeds, airline management felt that passengers preferred to stop en route and break up the trip. Hence, American and United

Yes, that's the Wright R-3350 engine that was causing all of the trouble for the Constellation. This hangar scene is most likely at Chicago Midway airport.

stopped at Chicago, Denver, or Dallas; TWA stopped at Albuquerque or Kansas City. And so, without the Rainbow to provide jet-speed travel, flying was still an all-day affair.

Each of the crosstown rivals had its attributes and detractions, but when it came to the bottom line, of the two original airframes, the Douglas airplane was the final winner. The Constellation looked shapely and beautiful, but all that beauty came at a price: The curvaceous fuselage and nacelles were more difficult to manufacture and took longer to assemble at the factory. The DC-6 had a virtually constant section cylindrical fuselage that provided consistent seating arrangements for its passengers and proved easy to stretch. But most important, it was all about the engines, again.

The Connie's Wright Duplex engines were temperamental and nettlesome to maintain and service. The DC-6 used the marvelous Double Wasp. Operational economics, which included maintenance and reliability, were measurably better for the DC-6 than the Constellation. So, although the Lockheed was a showstopper in looks, the solid and reliable Douglas airliner carried the day within the airline industry.

A candid shot shows busy American Airlines mechanics tuning a DC-6's R-2800 engine. This is one of the original DC-6s delivered with sleeping accommodations. Note the series of small sleeper berth windows above the main window line.

Flagship California NC90702 receives its motive power at the factory by mounting the propeller. (Photo Courtesy Boeing Co. via Jon Proctor with permission)

The birthplace of the DC Liners was Santa Monica, California. United DC-6s await delivery to the airline, and notice United's marketing ploy in the late 1940s to label its Mainliners by the airplane's airspeed. Hence, the DC-6 was a Mainliner 300, just as the DC-3 was known as a Mainliner 180. (Photo Courtesy Boeing Co. via Jon Proctor)

In Service

All of this advancing and pioneering by Douglas and Lockheed may very well have gotten ahead of itself once the proof of the pudding was at hand in everyday airline service. Both airplanes suffered difficulties, some normal teething for advanced airframes such as these, others were the result of enlarging the envelope, and still others were perhaps due to a bit of rapid zeal to meet the postwar market. Regardless, problems had to be addressed.

At Lockheed, the Constellation was having some major difficulties with its Wright engines and inflight fires thereof, the old problem from B-29 and C-69 days. Two problems were

Eastern's introduction of the Constellation was met with color photos and equally colorful accolades.

Typical early Connie interior with five-abreast seating. The airliner was wider in the midsection than the Douglas and could accommodate narrower seats. Notice the pillow/blanket overhead shelf.

An awesome piece of advertising illustration. Just look at that tray full of food! How could you not want to fly somewhere on a TWA Constellation?

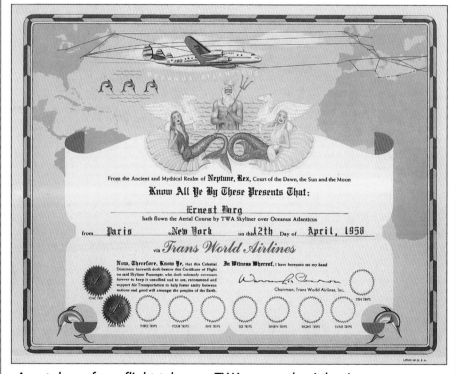

As a token of any flight taken on TWA across the Atlantic, passengers received this certificate that allowed for gold appliqués to be placed on each successive trip. PAA also had its version, recognizing that it was no small feat to fly internationally in 1948.

addressed by Lockheed and Wright on the engines, which included removal of the carburetors and installation of fuel injection, that allowed safe reliability to resume. The electrical system and its wiring layout also had a shortcoming that eventually caused a fatal crash. The airplane was prone to shorting and breakouts of fires within the system. This caused the CAA to ground the airplanes in the summer of 1946 until a remedy could be found, which eventually was, and the airliners resumed duty six weeks later.

Douglas had its own worries to address. A fatal crash in fall 1947 and a near-replication of the same two weeks later pointed to cabin fire as the culprit. Springing from the fuel venting system discharging raw gasoline into the cabin heater scoop under the fuselage, the entire aircraft became engulfed. Again, the CAA moved to ground a second modern new airliner until solutions could be found and fixes added. This process was considerably longer and kept the DC-6 fleet of 146 airframes idle or in production for some 140 days, until March 1948. As is always the case in aviation, progress requires tough lessons to be learned and impediments to be surmounted.

In 1947, Lockheed responded to the DC-6 threat by improving its original Constellation, the Model 049. In final form, the modified airplane, known as the Model 649, was some 50 percent reworked from the basic wartime airframe. Only Eastern bought this version whose improvements were almost immediately enfolded into yet another advanced version, the 749. This is the airplane the airlines ordered in quan-

A Speed Pack, which was winched up and into place beneath the airplane, also came with four tiny wheels built in so that it was fully movable and towable. (Photo Courtesy Gerald Balzer Collection)

The purpose-built DC-6A was either a freighter or a combination passenger/freight airplane. On the port-side fuselage were two large cargo doors that swung upward. The -6A was a stretched version of the standard DC-6 and was the basis for the military versions of this airplane. American had freight-only operations since the end of the war and through 707 days in the mid-1960s.

tity. It incorporated a longer range by an additional 1,000 miles, newer engine nacelles, increased cruising speed, greatly improved cabin soundproofing, systems upgrades, propellers with reversible pitch (begun on the DC-6), and better passenger accommodations. The point has often been made that this airplane was the real Constellation, matching precisely the airliner configuration and application originally envisioned by Lockheed prior to the war.

A very interesting and unique option developed for the Connie at Eastern's behest was an exterior luggage and cargo pod known as the Speedpak. Shaped like a canoe that conformed to the underside shape of the airplane, the capacity of the device was 395 cubic feet and 8,300 pounds of freight. Its use slowed the airplane by some 12 mph. Douglas matched this with its DC-6A version, which was cargo/cargo passenger with large doors in the fuselage and a strengthened floor. Several airlines plus the military used this airplane. Lockheed never sold a cargo/cargo convertible version of its basic Constellation model.

The average price of a new Connie was $850,000. All told, the first three models of the Lockheed airplane sold a total number of 226 aircraft. The DC-6/-6A totaled 249.

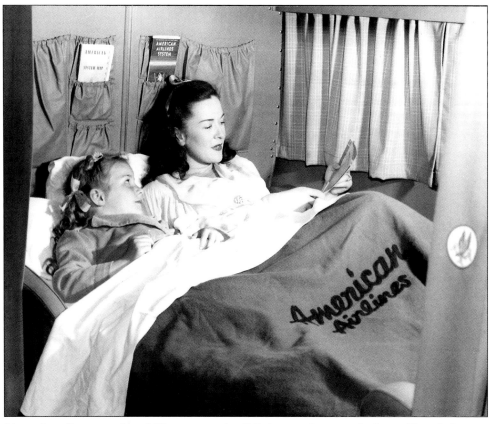

Noted earlier was the ability to use the DC-6 as a sleeper airplane. Here is how standard seats were turned into a bed. Note the script name on the blanket. (Photo Courtesy Museum of Flying Collection)

Going Somewhere

With the actors now on the stage, let's take a snapshot of long-range air travel in the summer travel season of 1948. PAA and TWA flew regular, daily service from the Northeast United States across the greater Atlantic Ocean to London, Paris, and on to the Mediterranean and Middle East. TWA finished in Bombay, India. If you flew that full route, you could be around the other side of the world in just over 36 flying hours' time. Shrinkage of the globe was now at hand. No longer hampered by water landing fields, the airlines were able at last to employ their land planes to ply the airways and extend American influence around the globe. So much had happened in just the few short years from 1940 that it scarcely seemed possible.

Of course, traveling by air long distances still required a way to sleep on the airplane. Douglas delivered many of the early DC-6s with the ubiquitous pull-down berths, including the small windows in the upper fuselage used as a tranquil salve to lull those star-gazing passengers to sleep. Lockheed chose to configure passenger seats as quasi beds unless the airline was operating the 749 series.

Several companies used the Pullman berth concept on those airplanes, including Qantas to Australia. The airplanes so modified also had the birth windows in the upper fuselage, small portholes just like the rest of the airplane. In either configuration, overnighters on both Douglas and Lockheed airliners were made to feel comfortable and at home as they traversed the many miles across oceans and continents. Breakfast in bed was a must and then off to the large dressing areas/lavatories to prepare for the day and destination's arrival.

It was quite an adventure, now made wholly routine by aeronautical innovation. Air travel had become accepted commonplace practice, thanks to the reliability of the engine technology at hand.

The competitive field was shaping up by the late 1940s: Douglas gave the world the DC-4 (C-54), which was met by Lockheed's 049 (C-69), which was then itself countered by the DC-6, only to be counterpunched by the 649/749 Constellation. Back and forth, like watching a tennis match, this magnificent dual did not stop with these airframes. Read on to see how these building-block aircraft were improved yet again and how that translated to the airlines and their customers.

Standard DC-6 working for Braniff in the early days. Note the factory-installed propellers and hubs. These particular types caused problems when they separated inflight. American almost lost an airplane due to these shortcomings. All airplanes later were retrofitted with Hamilton Standard props. (Photo Courtesy Boeing Co. via Jon Proctor)

Pearce

Hardware on the Wing

Wright R-3350 Duplex-Cyclone

The design of the Wright R-3350 was initiated in January 1936, shortly after development of the R-2600 began. Initially, Rudy Daub was in charge of the R-3350 engine design, but he was soon taken off the team to focus on another project, the R-2160 Tornado. The R-3350 and R-2600 were mostly developed in parallel and shared many components. The R-3350 used the same 6.125-inch bore and 6.3125-inch stroke as the R-2600. However, the R-3350 was composed of two rows of nine cylinders, creating an 18-cylinder radial engine. Total displacement for the R-3350 was 3,348 ci.

Also known as the Cyclone 18 or Duplex-Cyclone, the R-3350 had a three-piece crankcase. Many sources state that the engine's crankcase was initially made of forged aluminum, but some sources

contend that a steel crankcase was used from the start. Crankcases made from thin-wall steel forgings were in use by at least 1943, and if aluminum crankcases were ever used on the R-3350, it must have been on only the earliest engines. The R-3350 used a three-piece crankshaft that was secured together by clamping sleeves at the crankpins. From the start, the engine used silver-lead-indium bearings between the master connecting rods and the crankpins. The master rods were located in cylinders number-1 and -10.

The early R-3350 engines used cylinders developed from the R-2600, with a steel barrel shrunk and screwed into a cast-aluminum head. These cylinders had cooling fins machined into the cylinder barrel and a total cooling fin area of 2,360 square inches per cylinder. Mass-produced engines were switched to Wright's thin, aluminum W cooling fins that were rolled into grooves machined in the cylinder barrel. These cylinders had 3,838 square inches of cooling fin area. Late production engines used a forged aluminum head for added strength and improved cooling, both of which were found to be necessary. The cooling fins of the forged heads were cut into the head by cam-controlled gang saws. Unlike the cam rings of the R-2600, the R-3350's cam rings had four lobes per track and rotated at 1/8 crankshaft speed in the opposite direction of the crankshaft. All of the cylinders had one intake valve and one sodium-cooled exhaust valve.

To make the engine as small as possible, the nine cylinders on each row and the two rows of cylinders were separated by the minimum distance possible. This configuration cut the airflow to the rear cylinders and also made it necessary for the exhaust ports of the front row of cylinders to be positioned on the front side of the cylinder. However, overheating issues resulted in the exhaust port being relocated to the back side of front-row cylinders used on postwar versions of the engine. The cylinder rocker boxes had mounting points for the engine cowling. This helped ease engine installation and minimize the diameter of the cowling, but it also made cylinder changes much more difficult.

A Wright-designed single-stage two-speed supercharger was housed in a two-piece rear accessory section. Throughout its developmental history, the engine had many different propeller reduction gear ratios housed in various front sections. Common propeller gear reduction ratios of production R-3350s were 0.4375 or 0.35, and

The R-3350 A-series served mostly as prototype engines. Many changes were incorporated into subsequent versions of the engine. The R-3350 was the most powerful engine in (limited) production in the United States at the start of World War II. (Photo Courtesy Aircraft Engine Historical Society)

This rear-row cylinder from an R-3350 A-series engine had a cast-aluminum head and 42 cooling fins cut into the steel barrel. Efforts to improve cylinder cooling resulted in a switch to Wright's W cooling fins on later engines. (Photo Courtesy NACA/NASA C-1944-4556)

Only the R-3350 A- and B-series had forward-facing exhaust ports on the front-row cylinders. This configuration caused problems, but mass production took precedence over changing the design. B-series engines were installed in B-29s and C-69/L-049s. (Photo Courtesy Aircraft Engine Historical Society)

This rear-row cylinder from an R-3350 B-series engine had the W cooling fins, made of thin aluminum sheets rolled into grooves machined in the steel cylinder barrel. The use of thinner fins allowed approximately 52 to occupy the same area previously used for 42: a 24-percent increase. (Photo Courtesy NACA/NASA C-1944-5796)

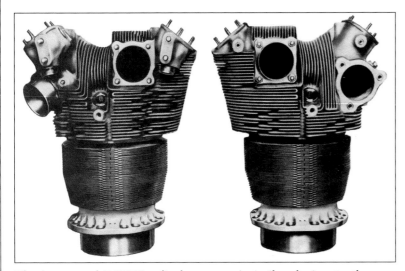

The improved R-3350 cylinders were a similar design to the R-2600's cylinders and still used cast-aluminum heads. There are 54 fins on the barrel, and they become wider closer to the head. The rear-row cylinder is on the left, and the front-row cylinder is on the right. Note that the exhaust port (square flange) for the front-row cylinder has been moved to the back of the cylinder. (Photo Courtesy Handbook of Service Instructions for Aircraft Engines Models R-3350-8, -14, -24, September 1945)

the reduction was achieved via planetary gears. The rear accessory section and the gear reduction housing were made of magnesium to save weight.

The R-3350 was first run in May 1937 and passed its military type test in January 1938. The R-3350 A-series engines initially had a 6.8:1 compression ratio and produced 1,800 hp at 2,200 rpm for takeoff and 1,500 hp at 2,100 rpm for normal operation using 100-octane fuel. The R-3350 A-series was 54.12 inches in diameter and weighed approximately 2,500 pounds.

The A-series entered limited production around March 1938. The engine experienced a protracted development as resources were allocated for the R-2600, and few customers were waiting to install R-3350s in their aircraft. By 1939, the only applications for the R-3350 were the Consolidated Model 31 (later XP4Y-1) experimental flying boat, the Douglas XB-19 experimental heavy bomber, and the Martin XPB2M-1 Mars flying boat, all of which were still under construction. The first flight of an R-3350 was in the Consolidated Model 31 on 5 May 1939.

On 29 January 1940, the Army Air Corps issued Request for Data R40-B, calling for a "Hemisphere Defensive Weapon" capable of delivering 2,000 pounds of bombs to the midpoint of a 5,333-mile unrefueled flight. Five bomber proposals were submitted, and all five proposed bombers were powered by the R-3350, the largest and most powerful air-cooled engine then under development in the United States. (Serious design work on the P&W R-4360 did not begin until November 1940, and with Allison focused on the V-1710, its V-3420 suffered from a perpetual lack of resources.) Suddenly, the R-3350 transitioned from a relatively low-priority experimental project of a private company to a strategic, high-priority engine for the military. The Boeing B-29 Superfortress and Consolidated B-32 Dominator were ultimately selected to fulfill Request for Data R40-B, and in April 1941, the Army Air Corps ordered the first batch of what amounted to more than 30,000 R-3350 engines to power these aircraft.

Wright had continued to develop the R-3350, and the B-series entered limited production in 1941. The B-series produced 2,200 hp at 2,800 rpm for takeoff and 2,000 hp at 2,400 rpm for normal operation. Early military versions of the engine had a 6.85:1 compression ratio and used 100-octane fuel. Commercial versions of the B-series entered production in 1945. These engines had a 6.5:1 compression ratio and used 100/130 PN fuel. The R-3350 B-series was 55.78 inches in diameter and weighed 2,668 pounds.

The Douglas XB-19 was first flown on 27 June 1941, and its R-3350 engines had constant cooling issues. The cowl flaps were required to remain open during all phases of flight to keep the engines from overheating. The drag created by the open cowl flaps slowed the XB-19 and reduced its range. The Martin XPB2M-1 Mars

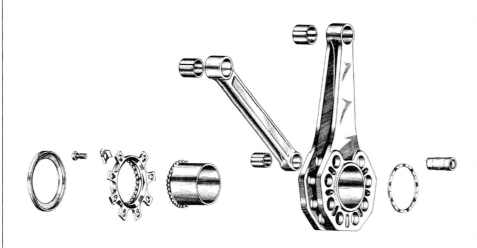

The master rod was one of the strongest components of a radial engine. However, the master rod occupying the number-10 cylinder (bottom of front row) could be bent if the engine was hydrolocked. Hydrolocking occurs when oil fills the cylinder after engine shutdown and is not drained (or cleared) before engine start. The incompressible nature of oil causes parts (such as a connecting rod) to fail as the piston moves in the oil-filled cylinder. (Photo Courtesy Overhaul Instructions R-3350-13, -18, -21, -23, -33, -35 Aircraft Engines, July 1943)

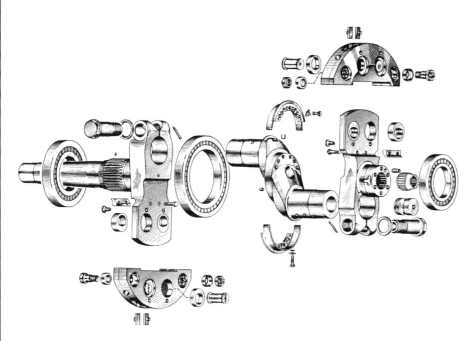

The R-3350's crankshaft and the R-2600's crankshaft shared a similar design. The front and rear sections were connected to the center section at the crankpins via clamping collars. (Photo Courtesy Overhaul Instructions R-3350-13, -18, -21, -23, -33, -35 Aircraft Engines, July 1943)

suffered an engine fire during an engine test on 5 December 1941. These issues were just an indicator of the trouble that lay ahead for the R-3350 engine. By the end of 1941, only around 15 R-3350s had been delivered to the Army Air Corps and the navy.

Wright and the rest of the aviation industry expanded as fast as possible to meet the influx of new orders as the United States armed itself in anticipation of the upcoming war. For R-3350 production, Wright built a new plant at Wood-Ridge, New Jersey, and converted its plant in Cincinnati, Ohio. In early 1942, the Chrysler Corporation was tasked with building the Dodge Chicago Aircraft Engine Plant in Chicago, Illinois, to produce R-3350 engines that were desperately needed to power B-29s and B-32s. These aircraft had the highest priority because the United States was now at war with Japan and possessed no aircraft capable of attacking that nation from land.

The Martin XPB2M-1 made its first flight on 23 June 1942. The Consolidated XB-32 flew for the first time on 7 September 1942. The Boeing XB-29's first flight followed two weeks later on 21 September. Around the same time, Lockheed's modified Ventura 4005, known as the *Vent-ellation,* flew for the first time. This aircraft had R-3350 engines in nacelles intended for the Constellation and a shortened nose to accommodate larger propellers. Thus, *Vent-ellation* was a combination of Ventura and Constellation. The aircraft's purpose was to refine the engine installation for the Constellation to minimize any issues.

With serious flight testing now underway and mass production of the R-3350 beginning to ramp up, a series of developmental issues with the engine arose. Gear reduction issues caused excessive wear and almost constant failure. Vibration issues caused the engine's propeller shafts to fracture or fail completely. Issues were encountered with the ignition system, and the engines experienced high oil consumption. Oil leaked past the piston rings and filled the lower cylinders after the engine shut down. With the cylinders hydrolocked, connecting rods were bent during startup if the oil was not drained. Fuel distribution issues of the carbureted engine resulted in the front row of cylinders running leaner than the rear row. This caused the richest cylinder to run 25 percent richer than the leanest cylinder and often resulted in violent backfires, particularly at takeoff power. Excessive fuel caused exhaust fires and fires after engine shutdown.

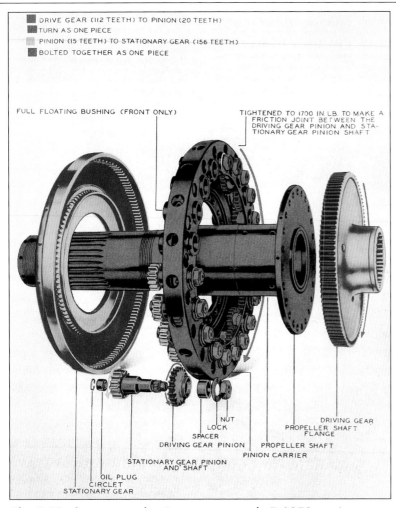

DRIVE GEAR (112 TEETH) TO PINION (20 TEETH)
TURN AS ONE PIECE
PINION (15 TEETH) TO STATIONARY GEAR (156 TEETH)
BOLTED TOGETHER AS ONE PIECE

FULL FLOATING BUSHING (FRONT ONLY)

TIGHTENED TO 1700 IN.LB. TO MAKE A FRICTION JOINT BETWEEN THE DRIVING GEAR PINION AND STATIONARY GEAR PINION SHAFT

DRIVING GEAR PROPELLER SHAFT FLANGE

PROPELLER SHAFT
PINION CARRIER
DRIVING GEAR PINION
SPACER
LOCK
NUT
STATIONARY GEAR PINION AND SHAFT
OIL PLUG
CIRCLET
STATIONARY GEAR

The 0.35 planetary reduction gear on early R-3350 engines caused many issues. The crankshaft drove the driving gear, which engaged 20 driving gear pinions (red teeth) housed in a pinion carrier. Each driving gear pinion was mounted on a shaft that had another pinion machined at its end. These pinions (yellow teeth) engaged a stationary gear mounted in the nose case. The propeller shaft was mounted to the pinion carrier, which turned at a reduced speed compared to the crankshaft. The gear reduction was redesigned for the C-series engines. (Photo Courtesy Overhaul Instructions R-3350-13, -18, -21, -23, -33, -35 Aircraft Engines, July 1943)

Cylinder cooling issues were also experienced; the lean mixture caused the front row of cylinders to overheat, and the lack of airflow caused the rear row of cylinders to overheat. Later in the engine's development, it was found that the upper cylinders of the second row ran 80 degrees Fahrenheit hotter than the cylinders in the front row. The forward exhaust port on the front row of cylinders only served to compound the overheating issues. The short exhaust stacks that joined the collector ring in the front of the cowling often cracked, releasing a flow of hot exhaust onto the cylinder and inside of the cowling. The exhaust valve in particular was prone to overheating and frequently failed, with its head breaking off into the cylinder.

By the end of 1942, flight testing of the non-turbocharged R-3350 engines in the XPB2M-1 and *Vent-ellation* were moving forward quite well. However, the twin-General Electric B-11 turbocharged R-3350s in the XB-32 and XB-29 were a different story. The XB-32 required 8 engine changes (12 total engines) before the aircraft reached 130 hours of flight time. The heavier and more complex XB-29 required 17 engine changes (21 total engines) before it reached 100 hours of flight time. No R-3350 had survived more than 20 hours of flight time on the B-29 or B-32 before an engine change was required.

A source of constant failure in the R-3350 was the engine's gear reduction for the propeller. Toward the end of 1942, it was discovered that many of the gear reduction pinions were misaligned during manufacture. Previously, Wright had tried nearly everything to resolve the issue but, apparently, did not think to check the parts for manufacturing accuracy. Until a long-term fix could be implemented, a single, highly-skilled machinist manufactured all of the pinion carriers on a nearly round-the-clock work shift.

On 18 February 1943, the second XB-29 prototype crashed into a building as the aircraft attempted to make an emergency landing at Boeing Field in Seattle, Washington. The crash killed all 11 people on the aircraft (2 bailed out but were too low for their parachutes to open) and 20 people on the ground. Two in-flight fires were experienced on the flight before the tragedy, and all XB-29 and R-3350-powered aircraft were subsequently grounded. Initially, the engines were blamed for the crash, and their previous difficulties certainly made that a valid assumption. However, the crash investigation revealed that it was a design flaw with the B-29's fuel system that had caused the fire; although, a number of engine deficiencies were noted as well.

Even though the crash was not caused by the R-3350, the engine now had a bad reputation, which resulted in more bad press for Wright. At the same time, R-2600 production issues were being investigated, and there were problems on the Curtiss side of the company with the SB2C Helldiver, C-46 Commando, and other aircraft projects. Wright persevered, and some of the R-3350's issues were solved, while others were minimized. Some 2,000 design changes were implemented between January 1943 and November 1943, but the fix for some of the issues could not be made due to the exigency of wartime.

As an example, Wright wanted to switch from cylinder heads made of cast aluminum to those made of forged aluminum. The forged heads were stronger and provided better cooling, which

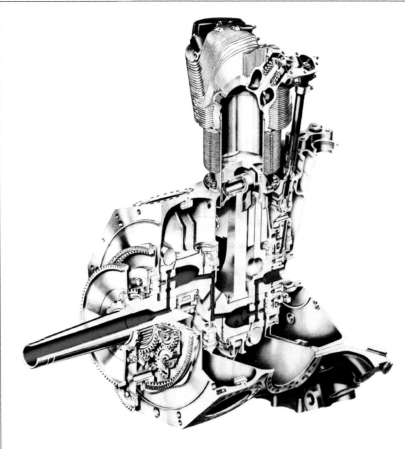

This cutaway of the R-3350 B-series engine shows a steel crankcase but still cast-aluminum cylinder heads. Note how the steel crankcase sections were bolted together internally between the cylinders. The progress of improving the engine was slow but steady. Oil distribution is highlighted in red. (Photo Courtesy Service Instructions R-3350-13, -18, -19, -21, -23, -35, -41 Aircraft Engines, September 1944)

ing the front cylinders to have a rear exhaust port would have necessitated changes to the entire exhaust system of the aircraft being built, delaying production. The B-29 was deemed too important to the war effort to slow production. By the end of 1943, only 1,001 R-3350 engines had been delivered, and 917 of those were delivered in 1943.

Although R-3350 production was somewhat delayed while various issues were resolved, the rush to mass production and the inability to make additional changes resulted in a bad situation. Engines prone to overheating and fire were encased in close-fitting cowlings attached to the wings of overloaded B-29s as they ran on the ground for extended periods before attempting to takeoff from a very hot environment. If a fire were to start, the magnesium gear reduction and accessory housings magnified the fire issue and decreased survivability. The tragic result was that many aircraft and crews were lost before techniques were developed to minimize the risk of fire and mitigate the heat-buildup of the engine.

While in flight, the increased drag of opening the cowl flaps to cool the engine slowed the B-29 and caused it to drop out of formation. In addition, a fully loaded B-29 could not maintain altitude

would have allowed for higher power outputs. However, the industry did not have the capacity for the change at the time. Also, Wright wanted to switch to direct fuel injection, but the required pumps were not available in the needed quantities early in the war. Chang-

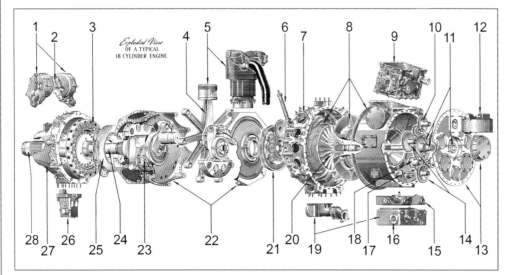

Exploded View OF A TYPICAL 18 CYLINDER ENGINE

*An exploded view of the R-3350. 1) ignition distributors; 2) torque indicator boost pump and distributor and governor drive; 3) propeller reduction gear; 4) rear connecting rod; 5) piston and rear cylinder; 6) valve pushrod tube; 7) supercharger front housing; 8) supercharger diffuser, diffuser shroud, and rear housing; 9) carburetor; 10) magneto drive; 11) accessory drive and starter shaft; 12) magneto; 13) supercharger rear cover; 14) rear accessory drives; 15) rear oil pump; 16) oil strainer; 17) tach and fuel pump drives; 18) generator drives; 19) rear oil sump; 20) impeller and shaft; 21) rear cam and vibration balancer; 22) crankcase main section; 23) front connecting rod; 24) crankshaft; 25) front cam and vibration balancer; 26) front oil pump and sump; 27) crankcase front section; and 28) propeller shaft. (Photo Courtesy Parts Catalog for Aircraft Engines Models R-3350-13, -21, -23, -23DW, *-23A, -35 and -37: May 1945)*

Although the engine was far from perfect, R-3350s in B-29s were constantly tasked to operate at their limits and under adverse conditions. The inconveniences of war forced engine repairs to be undertaken in less-than-ideal environments. Note the numerous engines scattered about in the background. (Photo Courtesy William Pearce Collection)

The exhaust stains from the Wright R-3350s show how hard the engines were run in this Lockheed L-649 Constellation. Eastern called their Connies "Silverliners," and the Ponce de Leon inaugurated service between Atlanta, Georgia, and San Juan, Puerto Rico, in June 1951. (Photo Courtesy William Pearce Collection)

with the cowl flaps open. As with ground handling and takeoff, procedures were eventually developed to manage the idiosyncrasies of the B-29 and its engines. Ultimately, B-29 bombers with their R-3350 engines proved to be a formidable weapon that brought the war to an end.

The story of problematic R-3350s in B-29s is widespread and true. However, more factors were at play than just the engine. Wright, a division of Curtiss-Wright, was unable to make completely independent decisions regarding engine development. This left both the R-2600 and the R-3350 underfunded and underdeveloped. Early testing of the engine in the Consolidated Model 31 and Douglas XB-19 did not approach the rigorous use the engine would see in the B-29. To save time while developing the B-29, no nacelle testing was done prior to the engine being installed on the prototype aircraft.

When all the pieces came together, it was found that the R-3350 in the B-29 had a short fuse when operating at its maximum output and that the role of the B-29 required the engine to operate at its maximum capacity and sometimes beyond maximum capacity. Changes could have been made to the engine and its installation to improve overall reliability, but as previously mentioned, the urgency of wartime production took priority over resolving all of the engine's

problems. The B-32 and the Constellation did not have as many engine issues as the B-29, but the engines in B-32s and Constellations were never pushed as hard as those in B-29s.

Some have said that more B-29s were lost because of the R-3350 than were shot down by Japanese fighters. Sadly, this statement is likely true. At least 414 B-29s were lost in the Pacific War. Of those, 147 were due to enemy action, and 267 were from other causes. Although no numbers have been found that break down exactly how many of the 267 lost were a direct result of engine issues, the raw numbers leave open the possibility that more B-29s were lost from engine trouble than fighters and all other enemy action.

In late 1943 and early 1944, an intensive program was undertaken by Wright and NACA to add anti-detonation (water) injection to the R-3350 engines installed in the B-29. The main appeal was that ADI would have provided an additional 400 hp per engine during the critical takeoff phase. However, NACA concluded that the injection would not lower exhaust valve seat or boss temperatures, and the valves were a weak point in the R-3350. In addition, a Wright representative dispatched to the Pacific to investigate issues with the R-3350 discovered that the engines were routinely run with cylinder head temperatures at 340 degrees C, well above the 290-degree C limit. The flight engineers had been told not to document any temperatures above 290 degrees Celsius.

Given those circumstances, Wright felt that ADI would lead to more problems rather than any solutions. The extra power provided by the injection would inevitably enable increased takeoff loads for the B-29, which would further degrade all flight operations once the injection was turned off. At the time, the R-3350 was being pushed to the ragged edge of its capabilities, and often beyond its capabilities. Any increases in the B-29's load would have pushed the engines beyond what they could endure. Wright decided to not move forward with ADI for the R-3350 until after the war, when some later engines were equipped with the system.

Many design features that Daub had originally incorporated into the R-3350 were removed after he was transferred to another Wright project, the R-2160 Tornado. As mentioned previously, Daub was not allowed to be associated with Wright during the war because of his ties to Germany. As engineers worked to resolve issues with the R-3350 engine, Daub's original designs were reincorporated into the engine, solving the problems. Had Daub stayed on the R-3350 engine project, or at least been allowed to consult on the engine, it may have had fewer issues, and the issues that it did have may have been resolved more quickly.

The R-3350 did not enter the civilian market until after the war, installed on the first Lockheed L-049 Constellations built for airliner use in 1945. Quickly, R-3350 engines were updated and incorporated changes that were desired during wartime, including forged aluminum cylinder heads, oil jets to cool the pistons, and fuel injection. The fuel injection was controlled by two pumps: One served the front row of cylinders and the other served the rear row of cylinders. This enabled different amounts of fuel to be provided to the different rows, curing the fuel distribution issues and aiding in cylinder cooling. Fuel was supplied to the injectors at 500 psi.

Continued R-3350 development resulted in the C-series, which entered production in 1947. These engines were essentially a complete redesign, incorporating some 6,274 design changes since the first R-3350 was built, such as forged aluminum cylinder heads. One of the more obvious changes was that the exhaust port for front-row cylinders was moved to the rear of the cylinder. This necessitated increasing the spacing between the cylinder rows by approximately 1.5 inches. The C-series had a 6.7:1 compression ratio and used 100/130 PN fuel. C-series engines had a takeoff rating of 2,700 hp at 2,900 rpm with anti-detonation injection and a normal rating of 2,300 hp at 2,600 rpm. The engine had a 55.62-inch diameter and weighed around 2,848 pounds. C-series engines were commercially available at the end of 1950.

Despite its troubled start, the Wright R-3350 Cyclone 18 was a popular engine for post–World War II commercial aviation. How-

Wright R-3350 (non-TC) Engines in Airliners				
Manufacturer	A/C Model	Engine Model	HP t/o	Number
Lockheed	C-69	711C18BA2	2,200	4
	C-69C	711C18BA4	2,200	4
	L-049	739C18BA2	2,200	4
		745C18BA1	2,200	4
	L-049, L-149	745C18BA3	2,200	4
	L-649, L-749	745C18BD1	2,500	4
	L-1049	956C18CA1	2,700	4
		975C18CB1	2,800	4

Major Applications of the Wright R-3350 Cyclone 18 Engine	
Boeing	B-29 Superfortress heavy bomber
Consolidated	B-32 Dominator heavy bomber
Douglas	A-1 Skyraider attack aircraft
	BTD Destroyer torpedo bomber
Lockheed	L-049 / L-149 Constellation airliner
	C-69 Constellation transport
	C-121 A Constellation transport
	L-649 Constellation airliner
	L-749 Constellation airliner
	L-1049 Super Constellation airliner
	P-2 Neptune maritime patrol aircraft
Martin	JRM Mars flying boat

ever, long engine life always required careful attention and smooth operation. The R-3350 was the least forgiving of the radial engines used in airliners. Further development led to the fuel-efficient R-3350 Turbo-Compound (D- and E-series engines) capable of 3,700 hp. Chrysler built 18,413 engines during World War II, and Chevrolet built 2,591 R-3350s in the 1950s. Production of the basic, non-compound R-3350 engine continued until around 1956. A total of 50,185 R-3350s were produced, and 38,231 (76.2 percent) of those were the basic, non-turbo-compound engine.

Pratt & Whitney R-2800 Double Wasp

At the same time that P&W's 14-cylinder R-2180 engine was about to enter production, the company had also designed a follow-on engine that was larger and more powerful. The new engine had two rows of nine cylinders. With a bore and stroke of around 5.625 inches and 5.8125 inches, respectively, the 18-cylinder engine displaced 2,600 ci. Construction of a P&W R-2600 prototype started in August 1936. When P&W learned of the Wright R-2600

Designed as a response to the Wright R-2600, the P&W R-2800 Double Wasp is arguably one of the finest piston aircraft engines ever built. It captured the proper balance of power and size while maintaining a very high degree of dependability. (Photo Courtesy William Pearce Collection)

in late 1936, it canceled production of the R-2180 and abandoned development of its R-2600.

P&W knew that it needed an engine more powerful than the Wright R-2600 and decided to use its 18-cylinder 2,600-ci engine as a starting point. However, the new engine would incorporate the larger 5.75-inch bore and longer 6.0-inch stroke cylinders of the R-2180. In a sense, P&W combined its R-2180 and R-2600 engine designs to create a new engine to compete with, and eventually surpass, the Wright R-2600. P&W's new engine displaced 2,804 ci and was known as the R-2800 Double Wasp.

The R-2800 engine program was led by Leonard (Luke) Hobbs, Andy Willgoos, Perry Pratt, and Wright Parkins. The R-2800 had an initial power goal of 1,800 hp, 300 hp more than the Wright R-2600. The engine followed the basic architecture of the R-1830, with a forged aluminum crankcase and cast magnesium used for the gear reduction and supercharger housings. However, the crankcase was made up of four pieces, with the middle section being split horizontally to be built up around the crankshaft. Common propeller gear reduction ratios were 0.35, 0.45, 0.40, and 0.50. Gear reduction was achieved through planetary gears. A single-stage supercharger was typically used, but later engines had two-stage supercharging. The supercharger was also available as a single- or two-speed unit. The two-stage two-speed supercharger consumed some 400 hp at take-off power on later R-2800 engines. P&W began developing its own superchargers in 1939, and they were incorporated into later R-2800 engines.

The engine's cylinders were composed of a steel barrel that was screwed and shrunk into a cast-aluminum cylinder head. For additional cooling, an aluminum muff with fine cooling fins was shrunk onto the cylinder barrel, and improved manufacturing techniques allowed for thinner, more numerous, and deeper cooling fins to be machined into the cylinder head. Each cylinder had one intake and one sodium-cooled exhaust valve in its hemispherical combustion chamber. The intake and exhaust ports were at the rear on all of the cylinders. The engine's cam rings had four lobes per track and turned at 1/8 crankshaft speed in the opposite direction of the crankshaft.

On the left (engine rear) of this R-2800 B-series are the large diffuser vanes (painted blue) that guided air to the supercharger's impeller. Note the thickness of the aluminum crankcase, especially the area supporting the center of the crankshaft. The large pinons of the planetary gear reduction are visible on the right (engine front). (Photo Courtesy William Pearce)

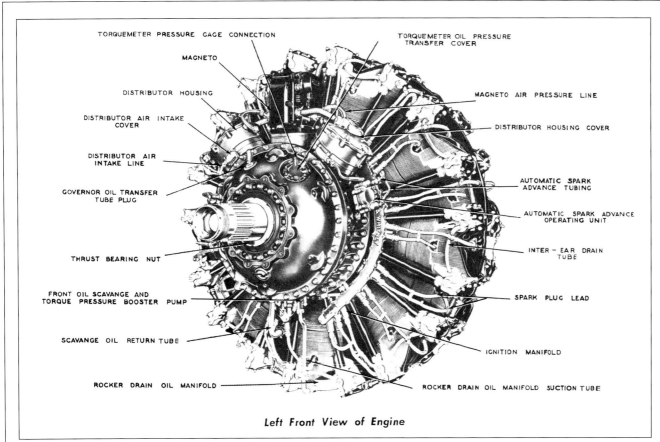

TORQUEMETER PRESSURE GAGE CONNECTION

TORQUEMETER OIL PRESSURE TRANSFER COVER

MAGNETO

DISTRIBUTOR HOUSING

DISTRIBUTOR AIR INTAKE COVER

MAGNETO AIR PRESSURE LINE

DISTRIBUTOR HOUSING COVER

DISTRIBUTOR AIR INTAKE LINE

GOVERNOR OIL TRANSFER TUBE PLUG

AUTOMATIC SPARK ADVANCE TUBING

AUTOMATIC SPARK ADVANCE OPERATING UNIT

THRUST BEARING NUT

INTER - EAR DRAIN TUBE

FRONT OIL SCAVANGE AND TORQUE PRESSURE BOOSTER PUMP

SPARK PLUG LEAD

SCAVANGE OIL RETURN TUBE

IGNITION MANIFOLD

ROCKER DRAIN OIL MANIFOLD

ROCKER DRAIN OIL MANIFOLD SUCTION TUBE

Left Front View of Engine

Various versions of the R-2800 powered everything from fighters to airliners, and it was even installed as a pusher in the Northrop XP-56. With high boost pressures, the R-2800 was coaxed into making 3,800 hp. However, maximum ratings for the basic engine never rose beyond 2,800 hp. (Photo Courtesy Overhaul Manual Double Wasp (R-2800) CA Engines, September 1948)

The R-2800 returned to the use of a single-piece master connecting rod and used a three-piece crankshaft. As with earlier P&W engines, the crankshaft was split at the crankpins. The splined crankpin sections were held together via a precision bolt. The master rods were located in cylinders number-8 and -13. The crankshaft was supported by silver-lead-indium bearings rather than the roller bearings that were used in previous P&W engines. The use of silver-lead-indium bearings for the master connecting rod continued.

P&W devised a unique system to deliver fuel to the R-2800's cylinders. From the injection carburetor at the rear of the engine, the metered fuel was fed into a tube that led to the intake side of the supercharger's impeller. The fuel flowed into a special device called a slinger ring. The slinger ring was positioned just before the impeller and had a series of small holes around its periphery. The fuel exited the holes in the slinger ring and was immediately atomized into the hot, high-velocity, pressurized air created by the impeller. The process homogenized the fuel into the incoming air charge and provided equal fuel distribution to the R-2800's 18 cylinders.

On 21 March 1937, P&W issued the order to begin construction on the first R-2800 (engine X-77), and development of the engine progressed rapidly. The engine was first run on 31 September 1937, and it had accumulated 100 hours on 18 November 1937.

The fuel slinger ring is positioned between the diffuser vanes and impeller. Fuel flowed from its numerous holes and was atomized as it mixed with the turbulent air being compressed by the impeller. The homogenous air and fuel mixture was then distributed to the cylinders. (Photo Courtesy William Pearce)

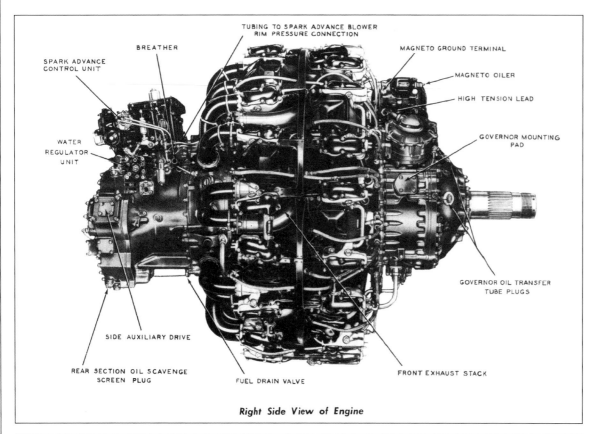

TUBING TO SPARK ADVANCE BLOWER
RIM PRESSURE CONNECTION

BREATHER

MAGNETO GROUND TERMINAL

SPARK ADVANCE
CONTROL UNIT

MAGNETO OILER

HIGH TENSION LEAD

WATER
REGULATOR
UNIT

GOVERNOR MOUNTING
PAD

GOVERNOR OIL TRANSFER
TUBE PLUGS

SIDE AUXILIARY DRIVE

REAR SECTION OIL SCAVENGE
SCREEN PLUG

FUEL DRAIN VALVE

FRONT EXHAUST STACK

Right Side View of Engine

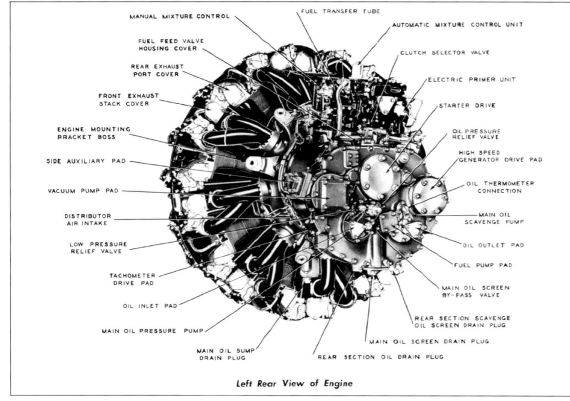

MANUAL MIXTURE CONTROL

FUEL TRANSFER TUBE

FUEL FEED VALVE
HOUSING COVER

AUTOMATIC MIXTURE CONTROL UNIT

CLUTCH SELECTOR VALVE

REAR EXHAUST
PORT COVER

ELECTRIC PRIMER UNIT

FRONT EXHAUST
STACK COVER

STARTER DRIVE

ENGINE MOUNTING
BRACKET BOSS

OIL PRESSURE
RELIEF VALVE

HIGH SPEED
GENERATOR DRIVE PAD

SIDE AUXILIARY PAD

OIL THERMOMETER
CONNECTION

VACUUM PUMP PAD

DISTRIBUTOR
AIR INTAKE

MAIN OIL
SCAVENGE PUMP

LOW PRESSURE
RELIEF VALVE

OIL OUTLET PAD

FUEL PUMP PAD

TACHOMETER
DRIVE PAD

MAIN OIL SCREEN
BY-PASS VALVE

OIL INLET PAD

REAR SECTION SCAVENGE
OIL SCREEN DRAIN PLUG

MAIN OIL PRESSURE PUMP

MAIN OIL SUMP
DRAIN PLUG

MAIN OIL SCREEN DRAIN PLUG

REAR SECTION OIL DRAIN PLUG

Left Rear View of Engine

With more than 125,000 built, the R-2800 was one of the most mass-produced aircraft engines of all time. The R-2800 was so outstanding that even after production had stopped, the engine was selected to power the Canadair CL-215 firebomber. (Photo Courtesy Overhaul Manual Double Wasp (R-2800) CA Engines, *September 1948)*

By January 1938, the R-2800 was producing 1,500 hp at 2,100 rpm, and 1,800 hp at 2,600 rpm was achieved in March 1938. The R-2800 test engines passed the 1,000-hour mark on 15 August 1938, and type tests were completed on 30 June 1939. The R-2800's first flight was in a Vultee V-11A (V-11T) on 12 July 1939.

The Vultee ground attack aircraft had been modified by P&W to serve as an engine test bed. After its modifications, it was designated V-11T.

The R-2800 entered limited production in 1939, but resolving a myriad of difficulties with vibrations and the engine's crankshaft delayed quantity production until early 1941. Initially, the engine had a 6.65:1 compression ratio and was rated for 1,850 hp at 2,600 rpm for takeoff and 1,500 hp at 2,400 rpm for normal operation using

From the supercharger, the intake manifolds were split to provide air to the front and rear cylinders. Note that the engine mounts angled toward the engine's center of gravity. (Photo Courtesy Overhaul Manual Double Wasp (R-2800) CA Engines, *September 1948)*

100-octane fuel. The R-2800 weighed around 2,350 pounds and had a 52-inch diameter.

As quantity production began, work was already well underway on an improved R-2800 B-series engine. Many minor changes were incorporated into the B-series, which had a takeoff rating of 2,000 hp at 2,700 rpm and a normal rating of 1,675 hp at 2,600 rpm on 100/130 PN fuel. The B-series had a 52.8-inch diameter and entered production in late 1941. As the United States rapidly expanded production of all military materiel to meet the demands of an impending war, producing the R-2800 engine in the quantities needed was difficult. Automotive manufacturers Ford, Chevrolet, and Nash-Kelvinator produced R-2800 engines during World War II, with Ford being the first and largest producer.

After leaving P&W on 27 June 1939, George Mead, P&W cofounder and former head engine designer, was appointed to the National Defense Advisory Commission in May 1940. Mead was tasked to oversee procurement of materiel for national defense, and he proved to be instrumental in enabling automotive manufacturers to produce aircraft and aircraft engines during World War II.

One of Mead's first undertakings was to visit P&W and discuss the licensed production of aircraft engines by automotive companies. P&W president, Frederick Rentschler, requested that the Ford Motor Company produce R-2800 engines. However, Ford was reluctant to take on the challenge. In August 1940, Mead and Hobbs visited Ford and convinced the company to move forward with aircraft engine production. The next day, 15 representatives from Ford started a two-week tour of the P&W plant. The Ford representatives mostly kept to themselves and did not seem particularly moved by anything they saw except for a table with rejected parts. The table contained parts that were perfect as far as the human eye was concerned, but with magnetic particle inspection, bits of iron could be seen aligned on microscopic defects. The rejected parts served to illustrate the manufacturing precision demanded for aircraft engines.

After evaluating what was needed to produce the engine, Ford representatives met with P&W representatives and declared that the R-2800 engine required manufacturing techniques that were beyond Ford's capabilities. However, if P&W would provide engineering drawings of the engine and blueprints of the P&W plant, Ford would have an identical plant built, outfitted with the same tooling and machinery, and would produce the R-2800 engine by duplicating P&W's R-2800 production. And that was exactly what Ford did. In just under a year, a plant was built, and production commenced, with the first Ford-built R-2800 engine completed in August 1941. It was, however, some time before the plant reached capacity.

Peak production occurred on 8 July 1944, with Ford building 186 R-2800 engines that day. If that rate of production were maintained, Ford alone could have produced 67,890 R-2800 engines a year. Ford developed a new process to manufacture high-quality cast-steel cylinder barrels for the R-2800 that could be built faster and cheaper than forged cylinder barrels. Another Ford innovation was to polish parts by tumbling them with small triangular stones, which saved an enormous amount of hand labor.

The ability of the United States to mass produce war materiel that was second to none went a long way to assure an Allied victory. Most of the factories that were quickly built at the beginning of the war did not achieve peak production until 1944, after the tide had already turned on the Axis powers. R-2800 production is just one example of hundreds that illustrate the incredible engineering and industrial prowess that helped win the war.

Continued development of the R-2800 engine resulted in the C-series, which was first run on 1 September 1940. The C-series was a complete redesign of the R-2800. Although the engine's basic configuration remained unchanged, virtually no parts were interchangeable between the C-series and earlier R-2800 engines. Cylinders for the C-series had forged aluminum heads and additional cooling fin area. The engine had a 6.75:1 compression ratio. Automatic spark advance was introduced, which adjusted the firing of the spark plugs to suit the engine's operating conditions. Internal components were strengthened, and the new crankshaft design enabled a return to the three-piece crankcase without a split center section.

The C-series crankshaft was still made up of three pieces and held together by a precision bolt passing through the center of the crankpin. However, the male and female splines had been discarded in favor of saw-tooth face splines joining the crankpin to the crankshaft cheek. New machining techniques made the face-splined crankshaft possible, and the design could handle more power than the A- and B-series crankshafts. The master rods for the C-series engines were moved to cylinders number-8 and -9 in an effort to lessen vibration.

The gear reduction was redesigned to handle the additional power produced by the engine. The gear reduction housing was now a two-piece casting but still made of magnesium. Common gear reduction ratios of the C-series were 0.45 and 0.4375. The 0.50 ratio used on earlier R-2800 engines had been found to excite vibrations and was discontinued. With 115/145 PN fuel, later C-series engines had a takeoff rating of 2,800 hp at 2,800 rpm with ADI and 2,500 hp without. The engine's normal rating was 1,900 hp at 2,600 rpm. The R-2800 C-series was 53 inches in diameter and weighed approximately 2,400 pounds.

The C-series entered production mid-1943 at a new plant built by P&W in Kansas City, Missouri. Production was delayed slightly at this plant because some of its tooling was diverted to Britain. Napier's H-24 Sabre aircraft engine used sleeve valves, and the company was in

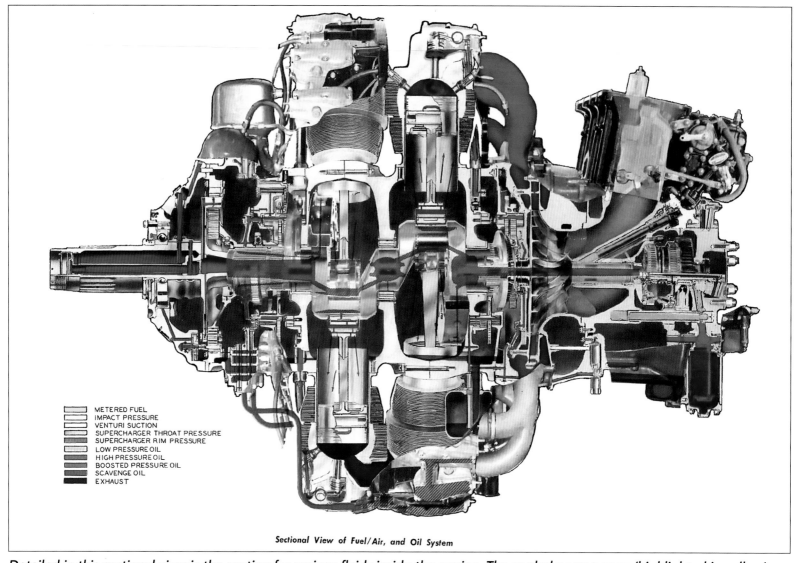

METERED FUEL
IMPACT PRESSURE
VENTURI SUCTION
SUPERCHARGER THROAT PRESSURE
SUPERCHARGER RIM PRESSURE
LOW PRESSURE OIL
HIGH PRESSURE OIL
BOOSTED PRESSURE OIL
SCAVENGE OIL
EXHAUST

Sectional View of Fuel/Air, and Oil System

Detailed in this sectional view is the routing for various fluids inside the engine. The angled passageway (highlighted in yellow) leading from the carburetor to the impeller provided fuel to the slinger ring, mounted just before the impeller. (Photo Courtesy Handbook Overhaul Instructions Models R-2800-50, -52W, -54, -99W, -103W Aircraft Engines, April 1956)

great need of centerless grinders to manufacture the sleeves. Grinders that were reserved for P&W's Kansas City plant were instead sent to Napier, and P&W had to wait for replacements. Such redistributions illustrate the urgency of war and the collaborative support of allies.

R-2800 D-series engines were based on the B-series but made to operate in pusher installations for the Northrop XP-56 Black Bullet. The E-series was based on the C-series but used a remote supercharger and a hydraulic coupling to drive accessories. The E-series was first run in November 1944 and entered production shortly thereafter. E-series engines had a takeoff rating of 2,300 hp at 2,800 rpm and a normal rating of 1,900 hp at 2,600 rpm. The R-2800 E-series was 53

inches in diameter.

As mentioned previously, the more powerful R-2800 engine replaced the Wright R-2600 in the Grumman F6F Hellcat and Curtiss C-46 Commando when development and production issues of the latter engine were encountered. The R-2800-powered F6F went on to become the most successful naval fighter of the war, destroying 5,223 enemy aircraft, more than any other allied fighter. The extra power of the R-2800 engines enabled the C-46 to fly its critical resupply missions from India to China, over the dreaded Himalayan Hump. The R-2800 engine combined with the Republic P-47 Thunderbolt airframe created one of the greatest fighters of the war and an aircraft

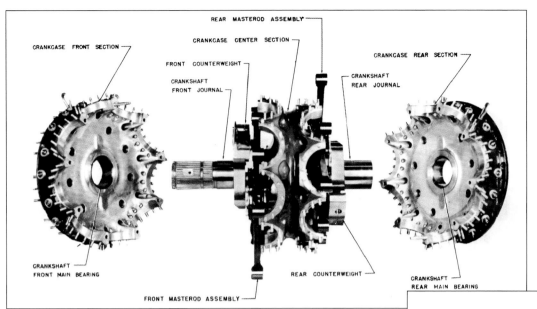

CRANKCASE FRONT SECTION
REAR MASTEROD ASSEMBLY
CRANKCASE CENTER SECTION
FRONT COUNTERWEIGHT
CRANKSHAFT FRONT JOURNAL
CRANKCASE REAR SECTION
CRANKSHAFT REAR JOURNAL
CRANKSHAFT FRONT MAIN BEARING
REAR COUNTERWEIGHT
CRANKSHAFT REAR MAIN BEARING
FRONT MASTEROD ASSEMBLY

The crankcase center section had a noticeable space between the front and rear cylinder rows, unlike the early R-3350s. Although it did make the engine longer, it also added support for the crankshaft. (Photo Courtesy Handbook Overhaul Instructions Models R-2800-50, -52W, -54, -99W, -103W Aircraft Engines, *April 1956)*

Being assembled here is an R-2800 B-series power section that is missing the front crankcase section. To the right, the crankcase's split center section can be seen on the assembled power section. In the foreground, the female splines of a crankshaft center section are visible. (Photo Courtesy LOC 2017694149)

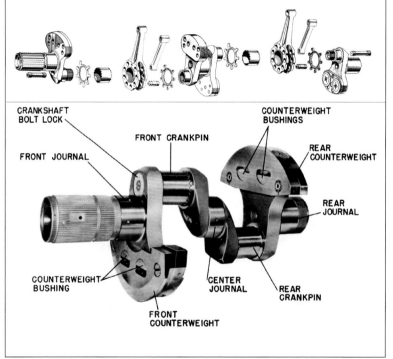

CRANKSHAFT BOLT LOCK
FRONT CRANKPIN
COUNTERWEIGHT BUSHINGS
FRONT JOURNAL
REAR COUNTERWEIGHT
REAR JOURNAL
COUNTERWEIGHT BUSHING
CENTER JOURNAL
REAR CRANKPIN
FRONT COUNTERWEIGHT

The top drawing is of the crankshaft design with coupling splines as used in A- and B-series engines. The lower image is of the improved crankshaft used in the C-series engines. The face splines are visible on the front crankpin. The saw-tooth face splines handled the additional power of the upgraded engines. (Photo Courtesy top: Overhaul Manual Double Wasp A and B Series Single Stage Engines, *July 1943; bottom:* Handbook Overhaul Instructions Models R-2800-50, -52W, -54, -99W, -103W Aircraft Engines, *April 1956)*

that continued to serve lesser air forces into the mid-1960s. A 2,800 hp R-2800 engine in the experimental (lightened and refined) XP-47J recorded a speed of 505 mph on 4 August 1944. This was the highest speed recorded in level flight by any propeller-driven aircraft during World War II.

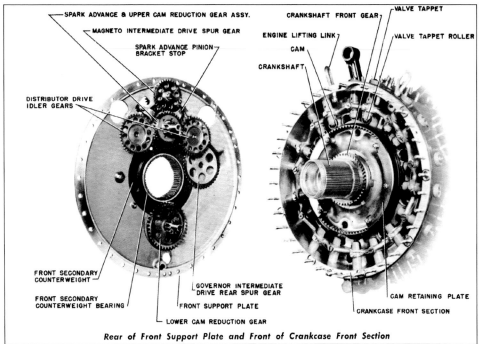

SPARK ADVANCE & UPPER CAM REDUCTION GEAR ASSY.
MAGNETO INTERMEDIATE DRIVE SPUR GEAR
SPARK ADVANCE PINION BRACKET STOP
DISTRIBUTOR DRIVE IDLER GEARS
FRONT SECONDARY COUNTERWEIGHT
FRONT SECONDARY COUNTERWEIGHT BEARING
LOWER CAM REDUCTION GEAR
GOVERNOR INTERMEDIATE DRIVE REAR SPUR GEAR
FRONT SUPPORT PLATE
CRANKSHAFT FRONT GEAR
ENGINE LIFTING LINK
CAM
CRANKSHAFT
VALVE TAPPET
VALVE TAPPET ROLLER
CAM RETAINING PLATE
CRANKCASE FRONT SECTION

Rear of Front Support Plate and Front of Crankcase Front Section

*This internal view of the R-2800's front section details the drives for the cam ring and ignition. The crankshaft extended through the front support plate and engaged the propeller gear reduction (not shown). (Photo Courtesy **Handbook Overhaul Instructions Models R-2800-50, -52W, -54, -99W, -103W Aircraft Engines,** April 1956)*

The excellence of the R-2800's design and the rivalry that existed within P&W are both illustrated in the story of Frank Walker's R-2800 C-series test engine. Walker was working on the ADI system for the R-2800 while the R-4360 was being developed. He took pride in making his R-2800 flawlessly outperform the larger engine; after all, the R-4360 usually achieved its power goal at the expense of a few broken parts. When the R-4360 hit 3,000 hp, Walker brought his R-2800 up to 3,200 hp. The R-4360 was then tested to 3,500 hp, so Walker matched it.

After the R-4360 made 3,800 hp, Walker again accepted the unassigned task of matching the larger engine with a 3,800-hp pull from his R-2800. Although Walker contemplated a 4,000-hp run with the R-2800, he decided it would be difficult to explain how he wrecked his engine during a self-assigned task of trying to match the performance of a much larger engine developed by the same company. In all his testing, Walker never blew up an engine.

Immediately after the close of World War II, R-2800 engines were made available for the commercial market. Nearly all of the R-2800 commercial engines were CA and CB engines, variants of the C-series. No two-stage supercharged engines were used for commercial aviation, but single- and two-speed supercharged engines were, some of which had ADI. The R-2800 was the first engine with ADI to be used by the airlines. It enabled the R-2800 to produce more than one horsepower per each pound of weight and allowed postwar airliners to take off with a maximum payload.

The postwar Convair twin-engine airliners and its military transport derivatives had perhaps the most unusual R-2800 installation. The Convair 110, 240, 340, and 440 used R-2800 engines in close-fitting cowlings. The cowlings did not

This cutaway shows the 0.50 propeller reduction gear used on an R-2800 B-series. In the center of the pinion carrier is a sun gear driven by the crankshaft. This gear engages the smaller side of the compound pinions held in the carrier. The larger side of the compound pinions engage a ring gear fixed in the nose section. The gear interaction caused the pinion carrier to rotate at a slower speed than the crankshaft. The propeller shaft is an integral part of the front pinion carrier. (Photo Courtesy William Pearce)

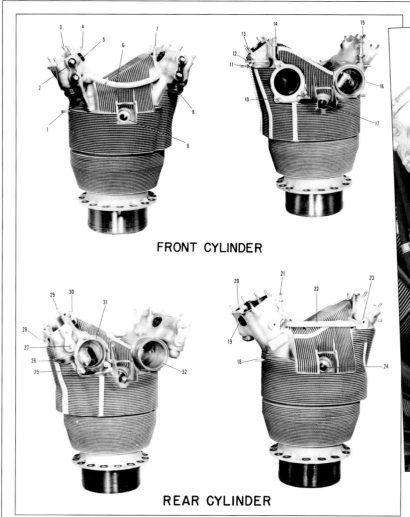

FRONT CYLINDER

REAR CYLINDER

1, 2, 4, 7, 11, 12, 14, 15, 18, 21, 28, and 30) Air baffle studs; 3, 13, 20, and 29) rocker box cover studs; 5) rocker shaft bushing; 6) inter-ear drain tube; 8) pushrod cover connectors; 9) spark plug inserts; 10) exhaust port studs; 16) intake pipe coupling; 17) thermocouple adapter; 19) rocker shaft pushing; 22) inter-ear drain tube; 23) baffle screw bushing; 24) spark plug insert; 25) exhaust port studs; 26) pushrod cover connectors; 27) baffle stud seals; 31) exhaust port liner; and 32) intake pipe coupling. (Photo Courtesy Handbook Overhaul Instructions Models R-2800-50, -52W, -54, -99W, -103W Aircraft Engines, April 1956)

P&W's cylinder designs never approached the number of small fins that Wright employed on later R-3350 engines. P&W used smaller cylinders than Wright, and the standard design of P&W's cylinders provided sufficient cooling for the R-2800. The lines across the cooling fins are beads of silicone that absorb vibrations and extend fin life. (Photo Courtesy William Pearce)

This front view of a front-row cylinder shows the leads to the spark plugs and that the piston was notched to allow clearance for the valves. The exhaust valve is on the right, and the intake is on the left. The tube at the top of the cylinder connecting the two valve rocker boxes equalized pressure and allowed oil to drain. (Photo Courtesy William Pearce)

have any cowl flaps. Rather, air taken in the front of the cowling was expelled through two ducts (or a single duct for the 440) at the rear of the cowling. The engine's exhaust was also routed to the duct, called an augmentor. Exhaust gases at 1,750 degrees F and traveling at 1,700 mph mixed with air from the cowling, expanding and accelerating the cooler air before it was expelled out of the augmentor. This outflow from the augmentor exhaust

This impressive view of an R-2800 in a Convair 240 illustrates an engine cowling designed with ease of maintenance in mind. The exhaust manifold can be seen entering the augmentor duct at the rear of the engine. The augmentor was also the outlet for engine cooling air. (Photo Courtesy Mike Machat Collection)

This side of a rear-row cylinder shows the intake valve actuation from the cam ring, through the pushrod, to the rocker arm, and finally to the open valve in its hardened seat. The hole in the cylinder reveals the steel cylinder barrel and its aluminum cooling muff. (Photo Courtesy William Pearce)

system created around 200 pounds of thrust for each engine and increased the aircraft's speed by 8 to 13 mph. Heat from the exhaust could be ducted through the wing's leading edge to prevent ice buildup, but the system proved to be insufficient under heavy icing conditions.

Two auxillary cooling doors were located above the engine and were used during ground runs. However, the augmentor system amplified engine noise in the passenger cabin, and overheating could occur if the aircraft was run on the ground with the wind coming from behind. The wind actually decreased airflow through the augmentor, which, in turn, decreased cooling air flowing through the cowling and cylinder cooling fins.

The R-2800 engine was also used in air racing and took the absolute speed record for piston-powered aircraft on 16 August 1969. That day, Darryl Greenamyer piloted his modified Grum-man F8F Bearcat *Conquest 1* to an average of 483.04 mph. Greenamyer broke the record set at 469.221 mph by German Fritz Wendel in the Messerschmitt Me 209 shortly before World War II.

P&W continued to develop the R-2800 and initiated design work on different engines using various combinations of turbochargers and turbo-compounding. Although these designs offered increased performance and decreased fuel consumption, they were also more complex and heavier than R-2800 engines in production. P&W proposed a turbo-compound R-2800 engine package for the Douglas DC-6B, but the proposal never found favor, and no turbo-compound R-2800 engines were built.

The engine in action on the wing of an Overseas National Airways DC-6A.

Licensed production of the R-2800 during World War II included 57,851 engines built by Ford and 4,282 engines built by Chevrolet. Nash-Kelvinator produced 17,012 engines during World War II and an additional 1,151 R-2800s during the Korean War in the 1950s. The R-2800 was in production until 1960, and a total of 125,443 were built. Interestingly, the last aircraft designed to use the R-2800 engine was the Canadair CL-215 amphibious fire bomber. The CL-215 was designed in 1965, years after the last R-2800 engine was built.

Many believe the P&W R-2800 Double Wasp was the best aircraft engine of World War II, and some believe it was the best piston aircraft engine of all time. The engine proved to be rugged and reliable, and it could take a fair amount of abuse. For the airlines, the R-2800's time between overhauls eventually reached 2,000 hours for twin-engine aircraft and 3,000 hours for four-engine aircraft. The engine was well liked by flight crews and mechanics, and is generally considered the most reliable of the big radial engines.

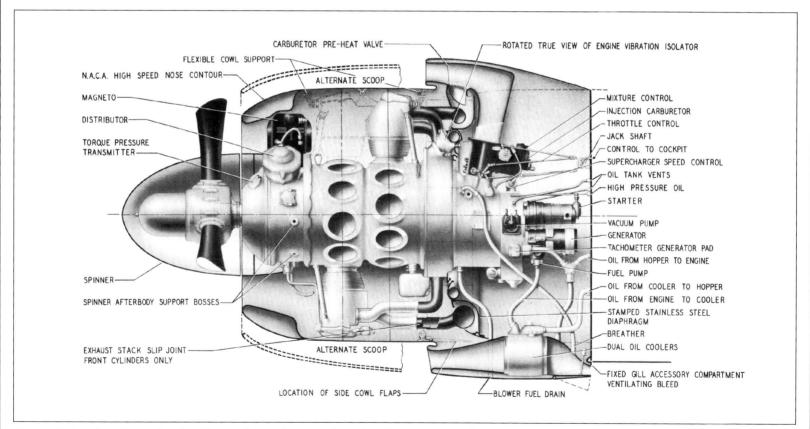

Beyond the aircraft engine itself, provisions for a multitude of accessories, ducts, and lines needed to be considered for installing the engine on the aircraft. All of this ancillary equipment was encased in a cowling that provided enough air to cool the engine but also created as little drag as possible. (Photo Courtesy Maintenance Manual Double Wasp (R-2800) CA Engines, October 1950)

Pratt & Whitney R-2800 Engines in Airliners				
Manufacturer	A/C Model	Engine Model	HP t/o	Number
Convair	240	CA15	2,400*	2
		CA18	2,400*	2
		CB3	2,400*	2
	340, 440	CB16	2,500*	2
	440	CB17	2,500*	2
Curtiss	C-46	R-2800-43	2,000	2
	C-46A, D, E, F, G	R-2800-75	2,000	2
Douglas	DC-6, A	CA15	2,400*	4
		CA18	2,400*	4
	DC-6A, B	CB16	2,500*	4
	DC-6A, B	CB17	2,500*	4
Martin	2-0-2	CA15	2,400*	2
		CA18	2,400*	2
	2-0-2A, 4-0-4	CB16	2,500*	2
	4-0-4	CB3	2,400*	2
* With ADI				

Major Applications of the P&W R-2800 Double Wasp Engine	
Bell	HSL anti-submarine warfare helicopter
Breguet	761/763/765 Deux-Ponts transport
Canadair	CL-215 amphibious firebomber
Chase	C-123 Provider transport
Consolidated	TBY Sea Wolf torpedo bomber
Convair	240 airliner
	340 airliner
	440 airliner
Curtiss	C-46 Commando transport
Douglas	A-26 Invader attack bomber
	DC-6 airliner
Fairchild	C-82 Packet transport
Grumman	AF Guardian anti-submarine aircraft
	F6F Hellcat fighter
	F7F Tigercat fighter
	F8F Bearcat fighter
Howard	500 executive transport
Lockheed	Ventura light bomber
Martin	2-O-2 airliner
	4-O-4 airliner
	B-26 Marauder medium bomber
	PBM-5 Mariner flying boat
North American	AJ (A-2) Savage medium bomber
Northrop	F-15 Reporter reconnaissance aircraft
	P-61 Black Widow night fighter
Republic	P-47 Thunderbolt fighter
Sikorsky	CH-37 Mojave transport helicopter
Sud-Ouest	SO.30 Bretagne airliner / transport
Vickers	Warwick (various versions)
Vought	F4U Corsair fighter

Power Developments During World War II

Around 1940, engineers and designers possessed all of the knowledge required to create high-power aircraft engines. Quality fuel was mixed with an incoming charge of air that had been pressurized by an efficient supercharger. This mixture was fed into a well-designed cylinder with a moderate compression ratio. The resulting power was gathered from all cylinders of a finely engineered engine and was transmitted through a robust propeller gear reduction. From the gear reduction, a large, metal, constant-speed propeller converted the engine's power into thrust and propelled the aircraft forward.

Possessing mechanical knowledge is only one part of creating a practical machine. Often, aircraft engine companies did not have the resources to experiment with engine concepts on a grand scale. Experimental development was expensive and slow. Engine companies often relied on the government to provide funds for research and development of new engine types.

Even though they did not have the funds to carry out their plans, engine designers were always thinking toward future developments. As World War II approached, the government began writing what were essentially blank checks to develop aircraft engines. Suddenly, the finances were available to put into practice all of the knowledge gained and all of the theories that had been speculated

about over the past decade. Wartime funding made it possible for cutting-edge concepts to be tested, developed, and produced. Money was available to develop manufacturing techniques that enabled the construction of more powerful engines and to cover the extensive testing needed to improve engine components, such as the supercharger. Small-batch fuel production became a thing of the past, and production of 100-octane, 100/130 PN, and 115/145 PN fuel began on a large scale.

Engine development that came out of World War II was simply a refinement of practices and theories that had been in place before the war. The influx of funds enabled development that would have taken 10 years during peacetime to be accomplished in 3 years. Although the aircraft piston engine greatly benefited from the intensive wartime development, it was ultimately the intensive development of the jet engine during the same period that made the piston engine obsolete.

Anti-Detonation Injection

ADI, typically a 50/50 mixture of water and methanol (methyl alcohol), could be injected into the intake of the engine to reduce the likelihood of detonation. The anti-detonation fluid mixed with the incoming air to reduce its temperature. The cooler charge also decreased the temperature of the cylinder. The result was that a more powerful incoming charge of increased density was taken into the cylinder. Because the temperature of the charge had dropped with the anti-detonation fluid, the charge had a higher resistance to detonation. The effect was a boost to the fuel's rating, giving 115/140 PN fuel the detonation resistance of 175 PN fuel. When ADI was activated, usually at a predetermined manifold air pressure, a de-enrichment valve was also activated to reduce fuel flow. The water and alcohol mixture replaced some of the fuel supplied to the cylinders; therefore, the fuel mixture could be leaned, which also produced more power.

With anti-detonation fluid being injected into the engine, more power could be produced without the risk of damaging the engine. However, only a limited supply of anti-detonation fluid could be carried, and the injection was typically only used during takeoff or for emergency power in military applications. A W at the end of the military engine designation signifies the engine has water (anti-detonation) injection, such as R-3350-26W.

In the United States, ADI originated with water injection only. In 1934, an Army Air Corps experiment was conducted at Wright Field, Ohio, in which water was injected into a P&W R-1340 Wasp engine running 90-octane fuel. The water-injection enabled the engine to tolerate a much higher intake manifold air pressure, and the engine's output was increased from 550 hp to 768 hp. However, the benefits of adding water-injection to aircraft engines was not seriously considered in the United States for a number of years.

Around 1940, P&W investigated how much water a running engine could ingest and still run. This line of experimentation was to ensure that engines could safely fly through heavy rainstorms. Engineers noticed that the engine's power could be increased as water was injected. Around 1941, P&W began to experiment with using water-injection to enhance an engine's performance. It was found that water-injection enabled the engine to produce 20 to 30 percent more power. Methanol was added to the water to prevent the fluid from freezing, which could easily occur in aircraft flying at high altitudes, and a 50/50 mixture produced the best results.

Starting in September 1942, P&W developed ADI with the intent of incorporating the system in production engines. P&W R-2800 engines with ADI went into production in 1943. Some airliners with P&W R-2800 and R-4360 engines and Wright R-3350 engines had ADI. Top modern-day air racers consume anti-detonation fluid at almost the same rate as fuel.

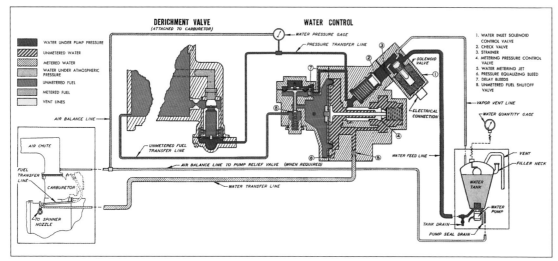

The Chandler-Evans Water Injection controller for the P&W R-2800 engine ran a system that reduced the chance of detonation and increased power. The water, or ADI fluid, replaced some of the fuel and was injected just before the supercharger impeller. (Photo Courtesy Overhaul Instructions for Chandler-Evans Water Injection Controls, 1951)

Torquemeter

A flight crew could only guess how much power an engine was producing until the torquemeter was developed. The torquemeter displayed a pressure readout from which engine power was derived. The torquemeter reading enabled the pilot or flight engineer to determine exactly how much power the engine was producing, so that other engine parameters could be adjusted for optimal performance.

As engine power was directed through the propeller gear reduction, a torque reaction occurred that could be measured by the exertion of a counter-acting force. In other words, a

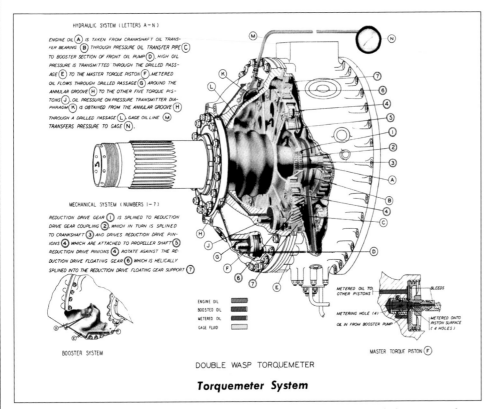

The torquemeter was a useful instrument that not only helped determine how much power an engine was producing but also indicated when an engine was failing. The torquemeter also helped determine the point of maximum engine efficiency. (Photo Courtesy **Overhaul Manual Double Wasp (R-2800) CA Engines,** *September 1948)*

pressure reading that was indicative of engine torque could be taken. The torque reading could be plugged into a formula to determine engine power. Depending on how the torquemeter gauge was calibrated, it read either a PSI value ready to be converted to engine power or the actual brake mean effective pressure of the engine.

Typically, brake mean effective pressure was displayed on the torquemeter. The reading was used to set power for takeoff, climb, and cruise. Once at altitude, the power was reduced from the climb setting, typically maximum except takeoff (METO) power, to read a specific, best-power value on the torquemeter. Then, the engine was leaned until a 10- to 12-percent drop registered on the torquemeter. This was the cruise setting for the P&W R-2800 and R-4360. For the Wright R-3350, the throttle was advanced after the 10-percent drop until the torquemeter read its original value. This "lean of peak" power setting minimized fuel consumption and maximized range. Close scrutiny of the torquemeter could also give some indication as to the overall health of the engine.

Practical and reliable torquemeters were introduced on the P&W R-2800 C-series and R-4360 and on the Wright R-3350 engines. They were also used on later models of the P&W R-1830 and R-2000 engines and on the Wright R-1820 and R-2600 engines.

The Flight Engineer

As aircraft systems became more complex, a flight engineer was added to the flight deck. This third crew member was responsible for managing the aircraft's engines and other systems during flight. The first aircraft with a flight engineer was the German Dornier Do X flying boat. First flown on 12 July 1929, the Do X was powered by 12 engines.

During the 1930s, US flying boats had flight mechanics who carried out duties similar to those that

First flown on 12 July 1929, the German 12-engine Dornier Do X flying boat was the first aircraft to have a dedicated flight engineer. The flight engineer's station was between two engine control panels, and each panel controlled six engines. (Photo Courtesy William Pearce Collection)

oil cooler doors. Fuel boost and transfer pumps and ADI were operated at various points during the flight. Two-speed superchargers needed to be shifted out of low gear and into high gear during the climb, and from high gear to low gear during the descent. Typically, two engines were shifted at a time. Most systems were manually controlled, and constant monitoring was required for the engine RPM, oil pressure and temperature, cylinder head and exhaust gas temperature, fuel pressure, and engine power as calculated by the torquemeter.

While all of this gauge-watching and lever- and switch-manipulation was underway, the aircraft still needed to be flown and navigated, so many of the above duties fell directly to the flight engineer. In many cases, it was the flight engineer, not the pilot, who manipulated the throttles for all phases of flight.

It fell on the flight engineer to anticipate the aircraft's power needs to ensure that the engines were always producing power and turning the propellers, rather than the propellers turning the engines, which is called windmilling. Radial engines in particular are designed to make power, not absorb power through windmilling.

Windmilling occurs when engine power is pulled back and the engine is turned (driven) by the propeller. This is most common during the descent and landing phases. Windmilling causes the reverse loading of engine components, which is particularly bad for the master rod bearings. The placement of the crankpin oil hole is critical to establish the hydrodynamic oil wedge upon which the master rod bearing floats. When the system is reverse-loaded (windmilled), the oil wedge is compromised, allowing metal-to-metal contact between the crankshaft and master rod bearings. This results in excessive wear or damage and drastically reduces the engine's life.

For the flight engineer, cruise flight did not always offer a lull between the high-workload periods of takeoff and landing. As the aircraft burned fuel and became lighter, engine settings needed to be adjusted to maintain the desired flight parameters. Changes in altitude, air density, and air temperature all required adjustments to the engine's systems to keep the flight going as scheduled. It was the flight engineer's job to continually monitor the instrumentation for any kind of engine trouble.

One of tools developed to monitor engine performance was the analyzer. Built by Bendix as the Electronic Ignition Analyzer and by Sperry as the Aircraft Engine Analyzer, the device was an oscilloscope that graphed the activity of the engine's magnetos, enabling its operator to detect, locate, and identify any abnormal ignition conditions. By fine-tuning the analyzer, the operator could view the action of a single spark plug displayed in wave form.

Certain issues inside the cylinder displayed on the analyzer in predictable ways. For example, a spark plug with a shorted

The flight engineer's station on a Lockheed L-1049 Constellation boasted a dizzying array of dials, levers, and switches. The throttle and mixture controls are in the middle. Built into the desktop is a circular viewport of the ignition analyzer. (Photo Courtesy Craig Kodera)

were eventually performed by flight engineers. The first US passenger transport with a flight engineer was the Boeing 307 Stratoliner, first flown on 31 December 1938. The 307 had four engines with constant-speed propellers, and the aircraft was pressurized. Each new innovation, such as constant-speed propellers and pressurization, added to the workload. The addition of a flight engineer relieved the pilot and copilot of many duties in the era before advanced automation, such as an autopilot. For the most part, all four-engine transports that followed the 307 required a flight engineer.

As engines and their associated systems became more complex, the pilot, copilot, and flight engineer had to manage engine throttles, fuel mixtures, propeller speed controls, and spark plug timing. They maintained engine temperatures by adjusting cowl flaps and

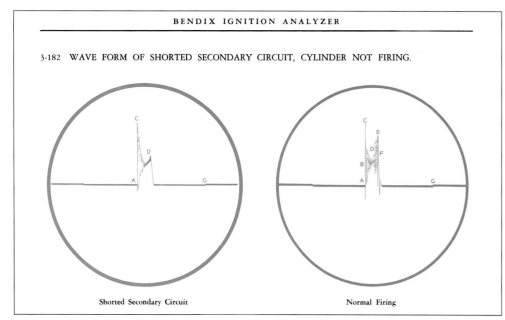

BENDIX IGNITION ANALYZER

3-182 WAVE FORM OF SHORTED SECONDARY CIRCUIT, CYLINDER NOT FIRING.

Shorted Secondary Circuit

Normal Firing

The ignition analyzer was an important tool used to determine the overall health of the engine. A variation of the wave form displayed on the scope (similar to this one) while monitoring a spark plug could indicate a failing cylinder. Under normal firing, A to C is the voltage spike when the spark plug first fires. The voltage drops to B almost instantly. The voltage varies as the plug continues to spark at D. The voltage increases to E, where the spark can no longer be maintained. The voltage suddenly drops to F, after which there are a series of voltage oscillations until the breaker points close at G. (Photo Courtesy Bendix Electronic Ignition Analyzer, February 1951)

An Eastern Air Lines flight engineer using the analyzer on board a Constellation. Always wear your hat in the cockpit when having a photo taken!

secondary indicated that the spark plug was not firing. The plug could be fouled or the lead wire could have disconnected, neither of which were a major concern. However, if both spark plugs in the same cylinder indicated a shorted secondary, it generally showed that the spark plug's ground electrode had been smashed into the center electrode. Such an event occurred when debris in the cylinder was slammed by the piston into the spark plugs. Such debris could be part of a failed piston or a valve that had broken.

Knowing there was a troubled cylinder, the flight engineer could use that engine's torquemeter reading to check for a drop in engine power, which confirmed a bad cylinder. The analyzer allowed the operator to accurately and quickly diagnose problems and limit damage to the engine.

Without these instruments, the poor performance of one cylinder on one engine was not noticed until a larger failure had occurred and resulted in additional damage or the complete destruction of the engine. A piston that was failing could seize in the cylinder. Once seized, the piston's connecting rod broke. A broken master rod immediately caused all of the other cylinders on the same row to fail in catastrophic ways. If a link rod broke, it flailed about its mount, damaging all other components with which it came into contact, causing other parts to fail within a short period of time.

Practices varied with engines and by airlines, but it was fairly typical for the flight engineer to check each spark plug every 30 minutes. There were 144 spark plugs on a DC-6, DC-7, and Constellation and 224 spark plugs on a Boeing 377. Although some of the work was tedious, catching a problem early could mean the difference between a minor inconvenience and a major catastrophe.

The more complex an engine, the more delicate operation it required. The Wright R-1820 and P&W R-1830 and R-2000 were very forgiving engines. The Wright R-2600 and P&W R-2800 required more care but still tolerated less-than-perfect operators. The Wright R-3350 and the P&W R-4360 required careful manipulation of the controls. The throttles needed to be handled smoothly and not too quickly. A sudden change in throttle position could damage an engine, or at least increase wear that would shorten its time between overhauls. The flight engineer made sure the engines were operating at their best and ensured their careful control.

From Swords to Plowshares and Back Again

By Mike Machat

It goes without saying that the development of military transports and commercial airliners have been intertwined since the beginning of aviation history. The very first Douglas airplane ever built, the robust six-passenger Cloudster, was morphed into the C-1 transport that conducted the first multiple inflight refuelings of another aircraft in 1929. It's amazing to think that history repeated itself 50 years later when the McDonnell Douglas DC-10 jetliner evolved into the KC-10 aerial tanker, 59 of which are still flying in operational service today. And perhaps there's no greater example of this phenomenon than the renowned Douglas DC-3 airliner being pressed into military service as the C-47 Skytrain, of which nearly 10,000 were built.

Although there are too many examples of this long-standing military/commercial relationship to highlight extensively in this book, two examples come to mind that epitomize this concept: the Boeing Stratocruiser and the Lockheed Super Constellation. Both evolved from military requirements to provide modernized, high-speed, heavy-lift payload capability to the US Army Air Forces at the end of World War II, and both became successful, if not legendary, airliners. For Lockheed's airplane, the mission came full circle from the army's original C-69 transport when the commercial L-1049 Super G Constellation was developed into the air force EC-121 and navy WV-2 Warning Star radar picket aircraft.

The story of the Stratocruiser's development, however, took an entirely different path, beginning with Boeing's then-mammoth B-29 Superfortress used in the Pacific Theater during World War II, and which carried the only two nuclear weapons ever detonated in wartime, putting an end to the war. With its Wright R-3350 radials replaced by 3,000-hp P&W R-4360s (and a taller tail for additional yaw control authority), the B-29 reemerged as the B-50. This airplane was then given a double-deck cargo-carrying fuselage to become the C-97 transport and eventually the KC-97 aerial tanker. The C-97's structural configuration formed the basis for the Stratocruiser.

Beyond the scope of this book are all the modern military aircraft that began life as commercial airliners. From navigational trainers to emergency medevac airlifters, or navy patrol aircraft to airborne flying command and control centers, airliners have earned their keep serving their country. Perhaps there could be no better example than the immaculate Boeing 747-200 that carries the president of the United States throughout the world today as a US Air Force VC-25!

The graceful lines of Lockheed's elegant Super G Constellation translate well to this early warning radar-equipped military version, a bare-metal US Navy WV-2 Warning Star. (Photo Courtesy Museum of Flying)

Looking like a promotional poster for the Wright R-3350 radial engine, this photo shows the Boeing B-29 super bomber of World War II as seen from an accompanying Stratofortress in formation. (Photo Courtesy Museum of Flying)

Boeing's Model 377 Stratocruiser represented the absolute zenith of airborne luxury in the postwar era, with sleeper berths, a lower-deck lounge, and first-class amenities throughout. (Photo Courtesy Mike Machat)

Chapter 7

SHRINKING THE ENVELOPE

Bearing in mind the postwar environment, most of the new equipment previously bought by the airlines of this country had been conscripted by the government to wage the conflict. The companies were, of course, reluctant to take back these war-weary aircraft for front-line service. It was a new age, so why not some new thinking about a new airplane designed to run the multitude of domestic short-haul and feeder routes operated by domestic air carriers?

Getting There by Air

Smaller Destinations

More people were now flying, thus the need for greater lift than the DC-3 could provide. Pressurized flying was the name of the game at this time as well, usually expected by the now-conditioned flying public (if only all the airlines felt this way!). And so, "America's Leading Airline"

A fine study of one of TWA's twin-engine Martin 202 aircraft after refurbishing. (Photo Courtesy Jon Proctor)

once again stepped up to the plate and put forth requests for a DC-3 replacement that could carry 40 passengers in maximum comfort and pressurized convenience.

Now Think Short Range

The major manufacturers in this country in 1945 responded to the concept of a modern twin-engine airliner with the following submissions.

- Boeing offered a 30-passenger high-wing design, which was not dissimilar to the later Fokker F.27. This was the Model 431.
- Lockheed put forth its diminutive Saturn Feederliner, designed for even lesser routes and so would carry only 14 fares, à la DC-2. Again, a high-wing design and too small; American said no.
- Curtiss submitted its CW-32, twin powered by R-3350 engines, but would seat only 32.
- Douglas came up with a unique design in its DC-8, a civil-

ianized version of its XB-42 Mixmaster twin-engine pusher bomber.

- Consolidated Vultee (Convair) showed its Model 110, which would tote 30 passengers.
- Martin chose the Model 202, an unpressurized airliner carrying 40 folks.

All the designs other than Lockheed's were circulated among the employees at American to gage their operational input. Amazingly, the winner of the preference poll was the DC-8. The airline's engineering department rejected the airplane, however, claiming the obvious expenses and operational difficulties if this layout were to enter daily service. The remaining two of the top three winners were the Convair and the Martin. Because the 202 was to be unpressurized, it was rejected at that point, leaving only one competitor standing: the Convair Model 110. Eventually, it too was subject to change by American.

Working with Convair, the airline stretched the 110, had the gull wing redesigned, and created a fairly drastic remodel of the airplane. The result was an airliner with two engines, designed to carry 40 passengers, and pressurized. The marketing people at the manufacturer suitably chose those two numbers to name the new airliner the 2-40. It also came to be known as the Convair Liner.

Here was an airplane that sat passengers in a quiet, nicely sized cabin, two abreast and four across. Pressurized of course, it innovated many airliner improvements: hot air taken from the exhaust pipes and used to heat and cool the interior and deice the airfoil leading edges. There were no more boots, but there was a drop-down ramp in the rear of the tail area, a passenger entry door ahead of the engines (it too had a folding "air stair"), reversible-pitch propellers, and exhaust that was channeled to the rear to augment engine thrust, kind of like a form of jet engine effect. Cruise speed was set at 270 mph, but the airplane could manage nearly 350 in a maximum effort. Just across the cabin from the main entry door was a coat closet, and two carry-on luggage racks were also available.

This airplane was designed to be self-contained and available for quick turnarounds at airports. With its cruising speed, businessmen could now fly to meetings in near-distant cities and return the same day. What a boon to American enterprise. (This demographic remains the strongest within the airline industry to this day and generates the most income for airlines.)

American was so pleased that in January 1946 it announced an order for 100 airplanes. At the time, this was the largest airliner order in history. The airline's future postwar fleet was now set: Convairs and DC-6s, and both would use the R-2800 engine. The bottom line was looking better and better thanks to all this new technology about to be implemented.

The 240 first flew in March of 1947 and entered service with American in June 1948. It was an instant success and customers loved it. It cannot be overstated that the world of airline flying from the passenger standpoint was almost totally different in the postwar climate. In fact, the change was so dramatic that when forced to fly on one of the new regional airlines, disparagingly known as puddle jumpers, whose fleets invariably consisted of hand-me-down DC-3s or wartime C-47s, audible disgruntled gasps could be heard from passengers. It was perceived to be a real step backward in this new age of modernism.

The last of the 240s was sold in 1950. By that point, a total of 176 airplanes had been sold to the airlines plus 390 various model types to the air force. The average cost for a new 240 was $495,000, not inexpensive but vital to building the new airline business of the postwar era. What a groundbreaking design it turned out to be. So, if you flew on American, United, Continental, or Western, chances were excellent that you would enjoy flying via Convair Liner.

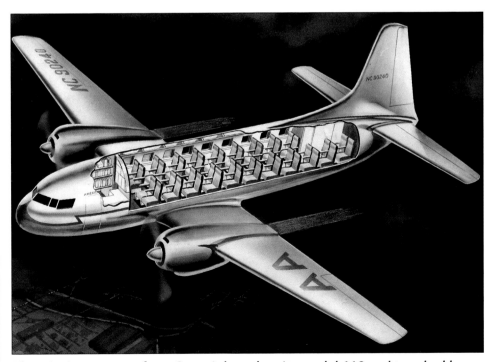

The winning concept from Convair based on its model 110 and tweaked by American Airlines. Note the artistic emphasis on the "jet" exhaust emanating from the aft end of the engine nacelles.

American's first Convair Liner photographed over the San Diego inland area in October 1947.

A close-up view of the innovative folding air stair found on the 240 and later series of Convair Liners. The engineers purposely kept the door in one piece and raised it above the ramp and doorway for rain protection.

Martin Answers

Meanwhile, across the continent from San Diego, Martin Aircraft of Baltimore, Maryland, was busy selling its own twin, the one rejected by American. The 202, even as an unpressurized airliner, found favor with airlines thinking more along the lines of DC-3 days and didn't mind the lack of higher altitude capability. But, if a carrier insisted on such, there was the Model 303, which was pressurized.

Although Martin lost the American order, the other airlines that chose Martin boosted the initial order book for both Martin airplanes higher than the one at Convair. Martin was well along with its two models early after the war and boasted a two-year lead time in service. This later evaporated due to redesign requirements mentioned below, but in those heady postwar days of "anything goes," all things were possible.

The 202 in many ways was a place-saver design for some companies that really wanted the 303 as their mainline airplane and when coupled with its early availability, clinched the deal. Martin's airliner entered service in October 1947, beating the 240 by nine months.

Despite all the later success at Convair, the reality of the postwar economic world was much different than the airline folks had imagined it would be. Namely, financing was difficult to acquire immediately after the war, and many of the orders Martin enjoyed in the first round of sales started to evaporate. Suddenly Braniff, Pennsylvania Central (Capital), and Panagra were no

Happy travelers disembark from a short Convair flight. American painted all its ground equipment the same orange as the airplane markings.

One of TWA's 202s, N93211, land at the Martin Aircraft factory in Baltimore, Maryland. Note that the production facilities still have wartime camouflage paint. (Photo Courtesy Jon Proctor)

The Martin 202 prototype, after the changes to the wing dihedral and huge dorsal fin installation for the vertical stabilizer. If only they had known about the weakness in the wing structure.

longer ordering. Then Northwest, which ordered both the 202 and 303, began to reduce its order total. But the biggest casualty in this financial war of attrition was United.

United Airlines believed fervently in the Martin 303. It was therefore a launch customer for the pressurized version of the basic airliner, spending $1 million of its own money to help influence the design. The upgraded version of the 202 flew a bit faster and could carry 36 passengers plus 4 additional in a rear cabin lounge. This airplane was to United as the 240 was to American and meant to be the workhorse of the fleet. Other airlines viewed it in the same context. Unfortunately, the previously bright world at Martin had been dimming when the 303 made its first flight as the problems with the very design of the 202/303 had begun.

Stability issues reared their ugly heads during flight testing of the 202 (also on the 303 because the airframes were identical), and this required a major rethinking of the airliner. The flat single-piece wing had to be kinked so as to include dihedral outboard of the engine nacelles. The vertical stabilizer was woefully skimpy and required an entire frontal dorsal to be added, basically doubling the surface area. The other rub in all this was that Martin had decided to start production of the original frozen design at the same time the prototypes were in flight test. Thus, all the airplanes being built on the production line had to be stopped and modified with the aerodynamic fixes. This is when all the time advantage mentioned earlier went up in smoke for Martin.

The Ultimate Martin

Because of these exigencies, United decided to review its commitment to the 303. It finally pulled the plug late in 1947 and literally got its $1.4 million deposit back from Martin. Convair was thus in a position to gloat and to coast. When TWA and United both approached the manufacturer about an improved 240 version with a longer wing to improve handling in the high-altitude airport regions of the western United States, Convair basically told the two to take it or leave it, referring to the existing Convair Liner.

In 1949, things began changing within the overall economic environment of the country, and Martin returned to the airplane concept it had evinced with its 303 model. No doubt a rude awakening for Convair, the improved twin from Baltimore was welcomed by TWA and Eastern, which ordered 65 of the type. Also looking seriously at this airplane were Braniff, National, Delta, Chicago & Southern, and Colonial. The competitive game was back on. The Martin 404 was on its way.

The improved airplane had many features incorporated from the learning curve of the 202. It now had a 2-foot fuselage stretch to accommodate a standard 40 passengers. The 404 was a decidedly new and improved version of the previous twin. The trouble was that it wouldn't be available for three more years. This caused TWA to lease an improved version of the 202, known as the 202A from Martin, 12 in all that were unsold airframes. TWA had good years of

Originally delivered in the standard bare-metal paint scheme, the 404 cut a proud vision in Eastern's colors. Missing in the 404 are the cockpit eyebrow windows found on its older sister. One can just discern the increased fuselage length of the improved, later airplane.

Ramp operations at Pittsburg, Pennsylvania, in November 1959. Note the open forward cargo/mail/crew door on the aircraft. (Photo Courtesy Jon Proctor)

service with this airplane, unlike the folks at Northwest, who seemed to be jinxed with their 202s.

We speak often in this book about teething problems as technology advances; most times they are normal and easily subdued, but sometimes they are drastic and deadly. This latter case was the story of Northwest and its Martin Liners. It seems that the rework of the wing design created a spar joining plate for those outboard wing panels that just wasn't robust enough. The plate failed, always in flight, and airplanes and occupants were lost. Two fatal accidents, coupled with a half-dozen other unrelated occurrences gave the 202 a tarnished reputation to the traveling public. It finally reached a point where Northwest pilots refused to fly 202s any longer.

After modifications to the wings, Northwest sold the airplanes on the secondary market. Because of the negative publicity, TWA quietly introduced its improved version of the airplane on its route system, awaiting the arrival of the 404. You have to wonder if apart from the lateness of the larger airplane's availability, perhaps the wing fatigue problems on the 202 may have permanently tainted Martin's reputation among the airlines. Regardless, the Eastern and TWA orders were the only ones other than the US Coast Guard for the 404.

The Martin Skyliner, as TWA named it, entered service with Eastern in January 1952. As was the case with the Convair Liner ahead of it, the 404 proved to be readily popular with the companies and the

You simply cannot take a photo of a Martin airliner without paying homage to the aft air stair. This airplane is currently flying and restored by the Mid Atlantic Aviation Museum in Reading, Pennsylvania. I have etched into my memory just this view of an Eastern airplane at Gainesville, Florida, in the very early 1960s, being viewed from the omnipresent chain link fence at the gate. (Photo Courtesy George Hamlin)

passengers who flew on it. Martin retained the rear-hinged loading ramp, a trademark of the airplane, and this feature was quite popular with passengers. As with Convair's arrangement at its entry area, the 404 had luggage stowage racks just at the top of the air stair as you entered the cabin. Similar to the 240, heating for the leading edges of all airfoils and air-conditioned air on the ground was provided by the pressurization system.

Convair Has the Last Word

As Douglas and Lockheed were engaged in their "Big Airplane" competition north of San Diego, Convair and Martin also were partaking in the twin-engine arena. Martin could not be allowed to best the Convair Liner with impunity and so, Consolidated got off its laurels and fought back with the improvements that TWA had asked for but was rebuked. This time United got the ball rolling and insisted, since it had abandoned Martin, that the 240's wing be lengthened for better hot and high airport performance as well as a stretch to the fuselage so as to carry more passengers. The new load would increase to 44 people. The improved airplane would be heavier and carry more fuel, using higher-horsepower R-2800 engines. Convair chose to name this model the 340.

One year before the start of 404 service, orders were announced for the bigger Convair Liner from United, Braniff, and Continental. Martin had been bested, although the 340 entered service in late 1952, so appearances favored the 404 for a while.

The 340 sold 209 airframes in its run, a testament to its money-making prowess as an airliner. United continued using the Convair into the early 1970s when it was finally supplanted by the Boeing 737. The 340 was one of those "just right" airplanes wherein the size, performance, and cost effectiveness (profitability) were just right. Convair was hitting all the proper notes and singing a beautiful melody. The 340 was well designed and rugged. It more than earned its keep.

There was some trouble on the horizon, however. Vickers introduced its Viscount turboprop, an airplane capable of outdoing the Convair Liners on all fronts. Convair decided to meet this challenge with a further improvement to the 340. Even better interior sound attenuation, new nacelle design to increase speed, higher-rated engines for increased speed, and a trailing-edge wing extension for better performance, the new airplane was called the 440 Metropolitan. It first flew in 1955, fairly late in the scheme of things. However, total orders for the new ship totaled a respectable 199 with the majority derived from foreign airlines, which in itself is respectable considering that the Viscount was an English airplane.

The 440 was a nice foil to the Viscount, and Convair succeeded in making the most of its twin-engine design over the 12 years it was active at San Diego. Of particular note in the gotcha category regarding the Martin/Convair competition was an order by the co-operator of the 404, Eastern Airlines, for 20 of the larger 440 aircraft. Martin had no further wish to beat its head against the Convair wall, and so Eastern simply had no choice but to mix its fleet types to accommodate growth and speed requirements in the short-haul category.

The larger Convair 340 twin with its apparent fuselage stretch aft of the wing.

By Judy Whiting James

Note: Judy's full story as an Eastern Airlines stewardess is found in chapters 8 and 9. We join her here in mid-career as she describes working the Martin 404.

The Martin 404, fondly known as "The Silver Falcon" (of The Great Silver Fleet), had a difficult rear boarding door to pull up and latch securely, especially by a wimpy-armed stewardess. The copilot usually had to come back and finish the job so we would not have to take a delay. That model was used primarily for short-haul puddle-jumper trips up the East Coast. At every stop en route, a full "coffee, tea, or bullion" had to be offered, and with only one stewardess that was quite exhausting, more so than on the much larger Constellation with 88-passenger capacity and three stewardesses. (If one of the three was a steward, he opted to work in the galley and also serve up the beverages there for the two women to serve in the cabin. That suited me just fine as I so enjoyed the personal contact with the passengers.)

Another fun time was on a Saturday morning "flight-seeing" trip with 44 Cub Scouts in a Martin 404 above Miami and Miami Beach. The morning began cloudy with scattered showers, so takeoff was delayed until the weather improved. And just what do 44 little boys do while waiting? They fill up on junk food from vending machines. That, added to the heat and humidity of South Florida, caused several of them to get sick before we even boarded. One sweet-faced little boy had gotten an instant crush on me and stuck close to me (candy included) and looked up so lovingly at me and said, "Gee, you're pretty" and then promptly threw-up all over my feet and uniform shoes.

After my own "emergency," we were soon ready to fly. Some had such tummy aches they did not get to do much sightseeing, plus it was still a bit bumpy at the lower altitude. If this was their introduction to flying, I would be willing to wager that they grew up taking the bus everywhere. So you can really understand the glamour associated with this job; all in a stew's day's work!

During spring training for the 1955 baseball season, I was called one morning to deadhead (fly an empty airplane to a destination) to Vero Beach to pick up the then–Brooklyn Dodgers to fly to Orlando to play against the then–Washington Senators. Orlando

A factory-staged plane full of passengers taking the rear stair of a Martin.

The efficient interior of Eastern's Silver Falcon. Note the immediate baggage stowage racks at the top of the air stair. Short range, multiple cities, and quick turnarounds were the life of a Martin Liner on the airways.

was in those days a sleepy little town, so having a major league team there in the spring was a very big deal. The Senators later became the Minnesota Twins, and the Dodgers became the Los Angeles Dodgers. Little did I or anyone else know that the Brooklyn Dodgers would win the 1955 World Series.

I can still see them all trudging up the rear-boarding stairs under the tail of the Martin. They were a very quiet group (make that worn out) and slept most of the way on the short trip. I had no idea there would be so many notable players on the team both then and in the near future. We never knew with whom we would be rubbing shoulders in the airline business, which made our time in the air that much more exciting.

Here is the pesky tail air stair that Judy complained of those years she flew for Eastern. (Photo Courtesy George Hamlin)

The ultimate Convair Liner in all its refined glory. Radar in the nose, longer wingspan, new engine nacelles, and several internal and cockpit upgrades made the 440 a true thoroughbred. A later sister ship, N9316, was the 1,000th Convair Liner built. (Photo Courtesy Jon Proctor)

The 440 joined Eastern's network in 1957. Eastern also had the distinction of receiving the 1,000th Convair twin produced, perhaps one more twist of the knife in Martin's side.

From the beginning, both competing twins were designed to accommodate turbine-propeller powerplants, which were the obvious "new, best thing" on the horizon (as witnessed by the Viscount challenge). Convair, however, became the only manufacturer to actually install varying types of turboprops on their in-service airframes, which extended their service lives many years beyond the time when most other piston-powered transports, large or small, had flown into the sunset. The regional airline industry enthusiastically embraced the turbine Convairs, and today, many of these airplanes continue to ply the skies in profitable bliss.

Chapter 8

THE BIG TIME

As the Lockheed-Douglas feud continued, the Santa Monica company introduced the DC-6B passenger version of the stretched A-model freighter. This longer and improved version of the basic DC-6 first entered service with United in February 1951. The B model was designed to fully counter the short-body Connie 749A series and represented the next punctuation mark in the ongoing escalation between the two manufacturers. It could seat 52 first class fares or up to 82 in coach configuration. It had Double Wasp engines from Pratt, at up to 2,400-hp with water injection. The DC-6B turned out to be the quintessential airliner from the standpoint of reliability, simplicity, and, therefore, profitability. Every aspect of this thoroughbred airplane's design seemed to synch and harmonize perfectly.

Getting There by Air

Pressurized Luxury

Turbo-compound engines throbbing, an American Airlines DC-7 wings its way westward to Los Angeles. Note later prop spinners.

What the DC-3 had done in 1936 to bring the airline business into the black on the ledger sheets and modernize/routinize the entire airline flying experience, the DC-6B accomplished in the early 1950s at the next, higher level. Travel by airplane was finally viable and enjoyable at the same time. There were 288 civilian airframes built, and as a testament to its superb balance of design features, the DC-6B was actually the last propliner to come off the Santa Monica line, outlasting the follow-on DC-7 series on 17 November 1958. The airplane type served with United Air Lines until 1976. In short, the Douglas became regarded as the touchstone of airline stability and economics.

Western Air Lines in Los Angeles in 1953 built its entire operation around the airplane; it was that perfect a fit for its route structure. Western always tried to bring "Big Airline" amenities to the medium-range stage lengths it flew, and the big DC-6 allowed

Brand spankin' new DC-6B on the ramp at Santa Monica. This airplane became the backbone for the airline industry and outlasted all other propliners in mainline airline use. (Photo Courtesy Mike Machat)

Western Air Lines used the DC-6B as its core airplane for years until the 707 and Electra arrived. This was the mid-1950s paint scheme, which fit this airplane like a glove. The light color on the top of all Western airplanes was not white, but rather an ivory or light tan. This photo places the airplane overhead Palmdale Air Force Production Facility in the upper desert of Southern California, just south of Edwards AFB.

Before Western's famed "The ONLY way to fly" slogan, DC-6B Champagne flights became popular on Western's routes.

NC6909C Star Of Tipperary seen flying over the Grand Canyon. TWA flew these basic stretch-model Connies starting in 1952. (Photo Courtesy Jon Proctor)

the company to create many novel high-level programs for its passengers.

Perhaps the most familiar for many years was the then-controversial serving of Champagne inflight. Begun with a certain amount of trepidation, the Californian Champagne Service was inaugurated in June 1954. What a hit it was! Drink and hors d'oeuvre imbibing took place in the First Class lounge located in the absolute aft section of the fuselage. Perhaps no other cutting-edge service option has ever been introduced by an airline (save inflight movies) that was then inexorably linked to a single carrier: "Western Air Lines, The O-o-o-only Way to Fly." Western's boss at the time, Terry Drinkwater, looked to the world to be a genius, and probably was.

And just as it had done with the Convair 440 order, Eastern, one of two stringent Constellation operators in this country, flew the DC-6B as well; it felt that good about the airplane. The Douglas workhorse found homes worldwide and cemented the maker's reputation for reliability.

All the same "new airplane" ideas simultaneously occurred to the thinkers in Burbank, and they too pulled at the fuselage of their airliner, adding just over 18 feet to its length. The cabin windows were finally brought into the modern age and made rectangular, new Wright engines were installed along with more fuel tankage, which increased both the speed and range of the airplane. New stronger and simpler construction methods were designed into the airplane, dropping the parts weight and allowing more airplane to emerge. More than 500 improvements were made to

CURTISS-WRIGHT TURBO-COMPOUND ENGINE

This art illustrates the temperamental Wright R-3350 Turbo Compound engine. The troublemaking Power Recovery Turbines (PRT) are at the rear of the engine and look like giant mushrooms.

the basic Connie. The new designation was the Model 1049 Super Constellation.

The single fly in the ointment at Lockheed was that the larger airplane was really meant to have the Wright turbo-compound engines with just that much more thrust than the standard R-3350s and, thus, allow the Super Connie to realize its inherent capabili-

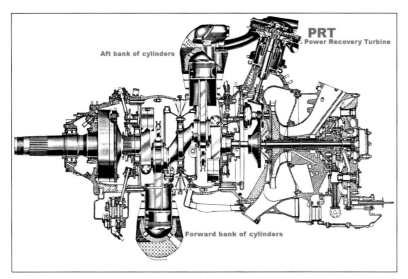

This cutaway of the R-3350TC illustrates the interconnection of the PRT to the cylinders.

The prototype DC-7 had the near-identical look of the fuselage of the DC-6B. This airplane went to American Airlines after certification.

ties. Without the proper engine, here is yet another example of no matter how great the airframe, success was not achieved. And so, the plain 1049 base model languished in the market place, only TWA and Eastern ordering the type. TWA, in fact, went back to Lockheed with its own couple of fixes/changes to the airplane and managed to boost the speed an extra 12 knots. The Super Constellation was still slower than the DC-6B, however.

First flying in 1951, the 1049 entered service later that year with Eastern and in Spring 1952 with TWA. This second round of contention went to Douglas and its DC-6B. But make no mistake in assuming that the Super Connie was not a large improvement over its short-body sister. Now Americans had two airliner plying the national skies that were luxurious, quiet, and roomy. The 1950s were settling in to be a lovely decade for all involved in airline transport.

Douglas Rides Its Wave

It was no secret that the turbo-compound engine was about to come to market for use in the airliner design business. As TWA worked with Lockheed to build an applicable airframe to take the engine, C. R. Smith at American, yet again, came forward to ask Don Douglas to incorporate the new engine into its line of airplanes and thus give American a competitive performance edge over its longtime adversary. The improvements promised by Wright with this iteration of its engine would now, theoretically, allow an airliner to fly nonstop across the United States, in both directions, and do so in less than eight hours, so just imagine the marketing

opportunities with this capability. The engineering staff at Douglas said it could match American's requirements, but management felt that the airline would be the only buyer of this faster airplane.

Accordingly, Douglas wanted to spend as little money as possible on this one-off experiment and so chose to use as much of the existing DC-6B platform as possible. The fuselage was stretched another 40 inches and heavier landing gear was included, along with the new engines of course, but that was about it. It had taken much arm twisting by Smith, but in the end Douglas said yes to perhaps its best customer and thus was born the next member on the crosstown rivalry team roster: the DC-7. American agreed to buy 25 plus 9 options. It later also ordered 24 DC-7B models. The airplane entered service on November 29, 1953, just six weeks later than TWA's quasi-transcontinental flights.

Much to the surprise of the always-conservative Donald Douglas, the seventh member of his commercial family was a hit. His basic model sold to United, Delta, and National as well, for a grand total of 105 units. There was more to glean from the airframe, and the second version of the DC-7 appeared as the B model. A logical extension of the airliner, this version was offered with nacelle-mounted saddle tanks for additional fuel. The range increased significantly, and PAA, South African, National, Panagra, and Eastern ordered the airplane. As a matter of fact, PAA operated the DC-7B nonstop from New York to London, quite a feat; South

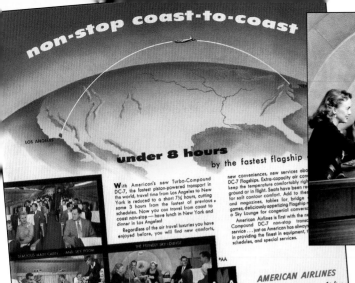

non-stop coast-to-coast

under 8 hours

by the fastest flagship

With American's new Turbo-Compound DC-7, the fastest piston-powered transport in the world, travel time from Los Angeles to New York is reduced to a short 7¾ hours, cutting some 3 hours from the fastest of previous schedules. Now you can travel from coast to coast non-stop—have lunch in New York and dinner in Los Angeles!

Regardless of the air travel luxuries you have enjoyed before, you will find new comforts, new conveniences, new services aboard DC-7 Flagships. Extra-capacity air conditioning keep the temperature comfortably right on ground or in flight. Seats have been redesigned for soft contour comfort. Add to these books and magazines, tables for bridge and other games, deliciously appetizing Flagship meals, a Sky Lounge for congenial conversation...

American Airlines' new Turbo-Compound DC-7 non-stop transcontinental service...just as American has always led in providing the finest in equipment, schedules, and special services.

AMERICAN AIRLINES
America's Leading Airline

This photo shows in full detail the scale and function of the lounge at the rear of the cabin. Notice the plexiglass partitions and the brass American eagle on the back wall.

From the introduction brochure of the American Airlines DC-7, this page spread shows the colors of a typical airline interior.

Here is that galley with coffee and water canisters galore and a lovely countertop space on which to construct the meals. Douglas called this area the airplane's buffet. (Photo Courtesy Boeing Co. via Jon Proctor with permission)

Included in one of those seatback information packs was a series of offers for models of the DC-7 so you could own a souvenir of the airplane. They made great gifts! I remember vividly filling out and mailing these forms and then eagerly awaiting the mail for the corrugated brown box with the model inside.

The forward section of American's cabin on the DC-7s was arranged as a club seating area and known as the Sky Room. The leather/vinyl/Naugahyde-like material pockets under the windows held the inflight information packets as well as magazines or other articles used by passengers. Note the pitch of the armrests on the seats designed by Douglas: You really sank into these luxurious chairs. (Photo Courtesy Boeing Co. via Jon Proctor with permission)

Standing in the galley of American's DC-7, which was located just inside the passenger entry door. There just seemed to be more room than what we have today. (Photo Courtesy Boeing Co. via Jon Proctor with permission)

With the success of the DC-7, Douglas improved the breed and gave the airline world the DC-7B. Here in Santa Monica is shown the saddle fuel tanks above the engine nacelles. Only a couple of airlines, however, took advantage of this range-enhancing feature. (Photo Courtesy Museum of Flying Collection)

The business end of the DC-7B powerplants. This is a PAA airplane posed for the camera at Santa Monica predelivery to the airline. Note the deicer boots on the leading edges of the blades as well as the streamlining prop hub spinners. (Photo Courtesy Boeing Co. via Jon Proctor with permission)

National loved its DC-7B airplanes. It had also flown the original DC-7, but the B was just right for the trips from New York to Miami, which were the airline's bread and butter. Still wearing the moniker "Airline Of The Stars" is airplane N6201B, the first of four for National. (Photo Courtesy Jon Proctor)

DECEMBER 21, 1953

There was a modicum of misery with the choice of the Wright turbo-compound engine that left much to be desired in the way of reliability. When it worked, however, the airplane sure flew fast.

African Airline flew service from Johannesburg to London via Rome.

Yes, Eastern was back at Douglas to buy an airplane that bested its Super Constellations, as mentioned earlier. Funny how things work out. The total ordered for all airlines was 112 of this later version. The airplane first entered service in June 1955. It seems that the anticipated orphan airliner had found a strong niche at the top of the food chain and flourished.

Standing on the Outfitting Ramp at Douglas, Eastern's first DC-7B is receiving final touches. The markings on this aircraft deserve some attention as the first iteration had the nose crescent painted red with blue pinstripes, just the opposite of the rest of the fleet of Eastern airliners. This was changed prior to delivery, and the spinners were painted red. Also, originally, these airplanes were delivered without radar. (Photo Courtesy Mike Machat)

How it would look in service, this Douglas photo provides a good look at the boarding ramp as well as the 72-inch-high passenger entry door. The Golden Falcon motif was used throughout the interior of the airplane in all the upholstery and fixtures as all things, plastics and wall coverings, etc., were royal blue and gold in color. (Photo Courtesy Mike Machat)

Meanwhile, Back at Burbank

Not taking well the chagrin caused by its basic 1049 models' power, speed, and transcontinental deficiencies, Lockheed fought back once more with its own turbo-compound version of the four-engine airliner. Upgrading the basic 1049 package, the 1049C made its first flight with the new engines just three months before the DC-7. The speed of the Connie now rested at nearly 330 mph, a nice increase but still slower than the competition. Notice how near the timing of the two airliners was. Eastern was the domestic airline to use the C model, the rest of the orders going overseas to airlines such as KLM, Qantas, and Trans Canada.

Unlike the basic DC-7 version, Lockheed sold only 48 examples. A further increase in structural strength allowed for higher gross weight and this yielded the 1049E Super Constellation. The two airplanes looked identical, and the heavier airplane garnered orders for 28 examples, solely from international carriers.

The DC-7B made it clear that the Connie needed more range. Lockheed's answer, as it had filled all the internal areas of the airplane and wing box with additional fuel tanks, was to go external. As had been developed for its military derivative 1049s (and what seems like just about every other airplane in their stable!), removable wingtip fuel pods were installed. So was born the distinctive 1049G model. The G could also have weather radar built on as an additional feature. This wise choice by TWA allowed it to label what looked like an entirely new airplane as the Super G Constellation. The aircraft turned out to be absolutely iconic, and as far as TWA was concerned, ubiquitous. All told, 102 airframes were constructed, making the G the most prolific version of the Constellation yet. Eastern and Trans Canada rounded out the North American airlines operating the G.

The new, longer legs of the Super G allowed TWA to fly nearly nonstop from New York to Paris, so combined with PAA's DC-7B service across the Atlantic, Europe was now easily within reach for the

Here is the quintessential Constellation, the one that everyone thinks of first when mentioning the triple-tailed airliner: the Super G. Looking for all the world like the Monogram model kit from 1956, the tip tanks are possibly the most intriguing aspect of this airplane. (Photo Courtesy Jon Proctor)

No radar nose and bare-metal fuselage distinguish the stretched-model 1049C for Eastern. (Photo Courtesy Jon Proctor)

Douglas placed its passenger lounge in the rear of the cabin, but with the taper of its fuselage the only viable location for such a meeting place in the Connie was in the center of the cabin. These were the days of wood paneling in airline interiors, and nobody did it better than Lockheed. Of special note is the Dreyfuss-created art mural along the sidewall. This photo is from a Northwest Airlines brochure, explaining the colors.

This photo is also of the lounge found aboard TWA's Super Gs. It is hors d'oeuvres time with champagne.

average passenger. More and better sound insulation was in place on the recent Connie models, making them real pleasures to fly aboard, especially over new, longer distances. And all the Connie operators used the passenger lounge areas provided aboard the Constellation.

TWA, in particular, pulled out all the stops and created a lovely oasis in the mid-cabin area of the airplane: juxtaposed sofas and seats and a table for cards or cocktails were all a part of the layout. A notable feature on these airplanes was special wall covering designed by Henry Dreyfuss and his design firm, appearing as art murals. Lavish champagne cocktails and hors d'oeuvres arranged with flowers on the center table lent a steamship-like flavor to flying international on TWA. Life in the air was once again reminiscent of the civilized flying boat era.

These elevated approaches to air transportation and life in general crowned the decade of the mid 20th century as perhaps the most glorious of all time periods. We wonder why we are nostalgic for those days. From an objective design viewpoint, it would seem that the designated winner of the style and luxury competition by this time in the process of airline passenger flying had switched to favor Lockheed.

Everyone asks, "Where was the lounge exactly?" Well, it was just ahead of the galley and passenger entry door. It is circled in this ad from Lockheed.

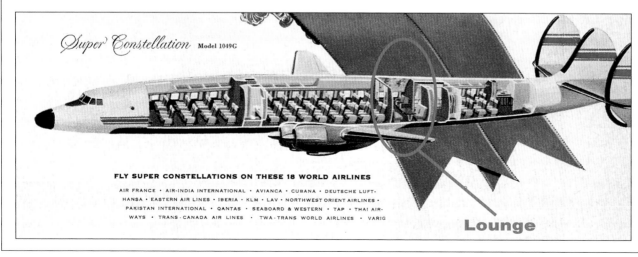

Super Constellation Model 1049G

FLY SUPER CONSTELLATIONS ON THESE 18 WORLD AIRLINES

AIR FRANCE • AIR-INDIA INTERNATIONAL • AVIANCA • CUBANA • DEUTSCHE LUFT-HANSA • EASTERN AIR LINES • IBERIA • KLM • LAV • NORTHWEST ORIENT AIRLINES • PAKISTAN INTERNATIONAL • QANTAS • SEABOARD & WESTERN • TAP • THAI AIRWAYS • TRANS-CANADA AIR LINES • TWA-TRANS WORLD AIRLINES • VARIG

Lounge

What's for dinner on a Super Connie? Well, at TWA it could be steak with mushroom demi-glace, potatoes au gratin, grapefruit salad, dinner roll, and a parfait of some sort with coffee, tea, or milk. And there is that bit of the bubbly just for luxury's sake.

While passengers enjoy dining their way across the continent, the front office of a Super Constellation is its usual frenetic self. This is the cockpit of a factory-fresh KLM 1049. Contrast this view with that of the Ford Trimotor in chapter 1. Pay attention to the engineer's panel and station here.

What had been sneaking up on everyone, inch by inch, was now about to come through the front door with a bang: the ability to fly nonstop from east to west in an airliner. Airlines were reluctant to do this, based on what was perhaps spurious passenger preference using the early DC-6s (it may very well have been due moreover to the CAA requirement that no pilot could fly more than eight hours a day and thus would require augmented crews, just as with international flying and, therefore, be expensive for the airlines), but now the urge and marketing tug became so great, and the airplane/engine available so viable, that they finally acquiesced and pulled the trigger.

American announced in 1953 that it would fly its new, yet-to-be-delivered DC-7, world's fastest piston engine airliner, nonstop, both directions, from New York to Los Angeles in just under eight hours, thus avoiding the regulatory restrictions of its flight crews. American's service was known as the Mercury, borrowing from earlier application on its DC-6 long hauls.

Hearing this news, TWA sprang into action. On 19 October 1953, it launched nonstop service between Los Angeles and New York with its 1049s and it called it "Ambassador Service." It was an overnight flight sequence, and the 1049s came equipped with the familiar pull-down bed system, unlike American's DC-7s, which had only a day-plane seating configuration. The westbound return took longer due to the prevailing winds over North America, so the total time aloft became a real sticking point with cockpit crews. Eight hours had been the regulatory maximum and that wasn't possible traveling west. The airplane had to stop in Chicago for a crew change. TWA flight crews negotiated with the company for overtime flying pay, and the CAA changed the maximum to accommodate the timing.

At American, however, the pilots didn't go quietly into the good night; after a stalemate with their management, they went on strike with the airline over this issue for 24 days. Although the westbound schedule was officially 7 hours 50 minutes, it was rare indeed to land in Los Angeles within that allotted time. Add to this the often normal situation of having to shut down one of the Wrights and the math became quite disparaging. Smith stepped back in and solved the dilemma for the recalcitrant DC-7 crews specifically, and flying resumed in August 1954.

As United received its DC-7s and American continued to grow its fleet, many more city pairs were turned into nonstop flights, including trips from San Francisco to New York and Los Angeles to Washington, DC. The airplane was indeed fast, at 359 mph, crossing the entire United States was now a one-day flying experience. A routine was established and laid the groundwork for the first jet service, also from coast to coast nonstop.

The DC-7 for American was an important stepping stone for this premier domestic carrier as it ushered in nonstop transcontinental flying. The airplane caught the fancy of the public, as seen here during the open house for the new terminal building at San Francisco. The moniker given to the San Francisco–to–New York service was, appropriately, the "Golden Gate." The young man in the photo is probably the station manager's son about to get into trouble messing with that sign. His father is at the base of the ramp explaining something to a visitor in line and is unaware of the impending mayhem. (Photo Courtesy Collection of SFO Museum)

One of the most majestic airliners ever built, the penultimate Constellation is just plain pretty. This is N7102C, which was initially named the *United States* in an out-of-character appellation because it was the inaugurating airplane for Super G service on 30 March 1955. (Photo Courtesy Jon Proctor)

Taking a breather at Gander, Newfoundland, in 1958, this H model was one of TWA's nine convertible freighter versions of the Connie, N6931C. (Photo Courtesy DAK with permission)

This is a rarely seen photo of a VARIG Brazil 1049 as it wings southward along the California coastline.

Here is a look at passenger terminal environments of the 1950s. Below is the then-new Tampa airport in 1952. I flew from this airport and terminal in 1962 on a National Airlines DC-8, which dwarfed the ramp area. On the left is the interior and ticket counter of United Air Lines at the San Francisco International Airport, the new building that supplanted the original Mills Field building shown in chapter 2. (Photo Courtesy Tamp Bay Library and USF Photo Collection with permission; from Collection of SFO Museum)

Pearce

Hardware on the Wing

R-3350 Turbo-Compound

In some cases, more than 50 percent of the energy available in fuel is vented out of the engine by the exhaust. The ability of turbochargers to harness some of this otherwise wasted energy is one of the reasons they have become popular. In 1942, Wright began to investigate other ways to recover energy from the exhaust and return it to the engine. Through its research, Wright developed the power recovery turbine (PRT) or blowdown turbine.

Exhaust gases were ducted through the PRT to drive a turbine wheel, much as they do in a turbocharger. But rather than boosting induction pressure like a turbocharger, the PRT was connected to the crankshaft via a stainless-steel shaft, gear train, and fluid coupling. Power from the exhaust was recovered and fed back into the engine. An engine equipped with PRTs was a turbo-compound engine, meaning its output was the combination of power derived

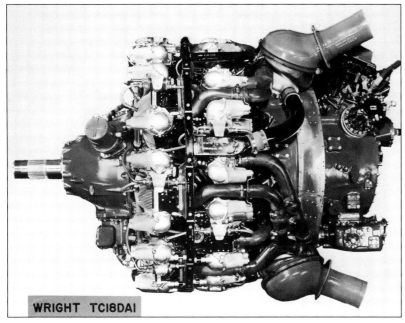

The Wright R-3350 turbo-compound engine series was perhaps the pinnacle of large piston aircraft engine development. These engines benefited from all of Wright's R-3350 experience and broke new ground by feeding power recovered from the exhaust back to the crankshaft. (Photo Courtesy Aircraft Engine Historical Society)

from the internal combustion of fuel and power recovered from the exhaust turbine.

Turbo-compounding should not be confused with a compound turbo system, in which two turbochargers are used in series, with a smaller, first-stage unit feeding a larger, second-stage unit. The Wright turbo-compound engines did not use turbochargers. In the 1940s, Wright's research indicated that turbochargers and their associated ducting weighed 1.3 pounds for every horsepower they recovered; PRTs weighed 0.9 pound for every horsepower they recovered.

PRTs did not create any back pressure and were shown to recover more power than turbochargers, especially at lower altitudes, with their advantage tapering off at around 30,000 feet. Another benefit of PRTs was that their inclusion did not add any controls or instruments for the pilots or flight engineer to manage. Also, a turbocharger could still be added, powered by the exhaust gases either before they entered or after they left the PRT. Wright referred to a turbo-compound and turbocharged engine as a double-compound engine and thought it would be ideal for high altitudes. However, Wright did not build any double-compound engines.

In January 1946, Wright began experimental tests sponsored by the US Navy. For these tests, Wright installed a single PRT on an R-1820 engine. The tests indicated the potential of the concept, and experimentation shifted to an R-3350 engine by May 1947. The first R-3350 Turbo-Compound engine used six PRTs, each directly collecting exhaust from three cylinders. Tests soon indicated that two exhaust pipes could be joined with no reduction in performance. The

Although it has seen better days, this R-3350 turbo-compound cutaway reveals the engine's complexity. A PRT can be seen on the left side. The turbo-compound had forged aluminum heads, a new crankshaft, and the latest gear reduction. (Photo Courtesy William Pearce)

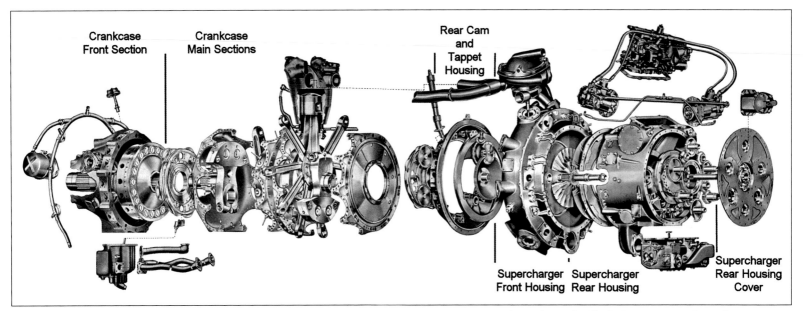

Crankcase Front Section

Crankcase Main Sections

Rear Cam and Tappet Housing

Supercharger Front Housing

Supercharger Rear Housing

Supercharger Rear Housing Cover

Illustrating all the various sections of the turbo-compound engine, it is easy to see that about half of the engine's length was taken up by the PRT section, supercharger, and accessory section. (Photo Courtesy Aircraft Engine Historical Society)

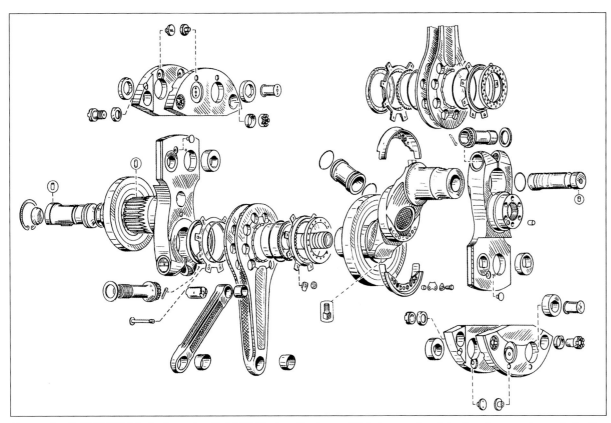

From the R-975 Whirlwind to the R-3350 Turbo Cyclone, Wright stuck with the clamped crankshaft design. The splined opening in the rear crankshaft section received the shaft that led to the PRTs. (Photo Courtesy Wright Aircraft Engines Series TC18DA Parts Catalog, March 1954)

exhaust manifold was bulged at the joint, and the incoming pipe extended a short distance into the manifold. This prevented backflow and aided scavenging. With this configuration, only three PRTs were needed, with each collecting exhaust from six cylinders. Each of the PRT's three inlets took in exhaust from two paired cylinders. Reducing the number of PRTs from six to three resulted in a 200-pound decrease in engine weight.

The R-3350 Turbo-Compound, or Turbo-Cyclone 18, was the pinnacle of piston aircraft engines. The engine used the most modern of the R-3350s, with a forged steel crankcase, forged aluminum cylinder heads, and fuel injection. A new section was added between the crankcase and the supercharger hous-

A display turbo-compound engine complete with chromed and polished parts. Note the supercharger's large impeller and the turbine wheel of the cutaway PRT (top mounted). Although it may be difficult to see, the fluid coupling has been cut away to reveal its runner. (Photo Courtesy William Pearce Collection)

ing that increased the engine's length by 11 inches. Mounted to this new section were the three PRTs. When viewed from the rear of the engine, the PRTs were mounted in 120-degree intervals at the three, seven, and eleven o'clock positions. Exhaust gases at around 1,500 degrees Fahrenheit and 200 psi entered the PRT traveling at 2,200 feet per second. The PRT's turbine wheel was 11.75 inches in diameter and spun at approximately 19,000 rpm at takeoff power and 16,000 rpm at cruise power. At takeoff power, each PRT returned around 170 hp to the crankshaft.

The turbo-compound had a strengthened crankshaft that weighed 157 pounds, 24 pounds more than a standard R-3350 crankshaft. The cylinder was improved, and additional cooling fins were added, bringing the total cooling fin area to 5,401 square inches per cylinder. The changes increased the cylinder's weight by 6 pounds, giving a new weight of 53.7 pounds for each cylinder. A sodium-cooled intake valve joined the sodium-cooled

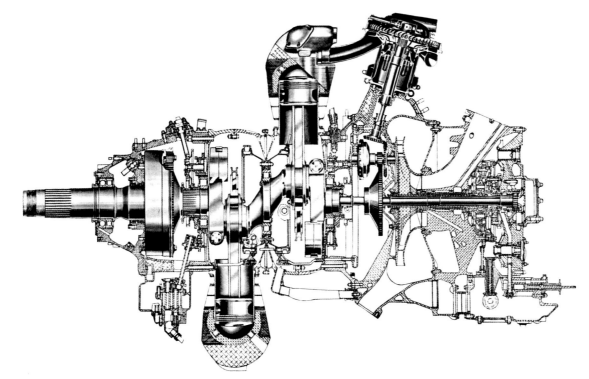

The three PRTs, spaced 120 degrees apart, fed power back to the engine's crankshaft via a gear mounted to an extension shaft. A separate shaft coupled to the rear of that same gear powered the supercharger. (Photo Courtesy Aircraft Engine Historical Society)

The final R-3350 cylinders were perhaps the most complex used on any radial engine. The cooling fins became progressively longer toward the top of the cylinder. This was an effort to maximize cooling and take advantage of the angled space between two cylinders on the same row while still providing sufficient airflow to the second row of cylinders. (Photo Courtesy William Pearce)

The roller tappet can be seen to the left of this rear-row cylinder. The large intake valve is in its hardened seat, and the piston has been dished for valve clearance. The steel cylinder barrel extends a fair distance into the crankcase. Note the relatively thin steel crankcase. (Photo Courtesy William Pearce)

exhaust valve already used on R-3350 engines. The engine's compression ratio was 6.7:1.

In August 1947, the navy ordered the first R-3350 turbo-compound engine, which was the D-series of the R-3350 line. The engine passed a 50-hour flight approval test in October 1949 and was test flown in the nose of a modified B-17 later that year. With no serious difficulties encountered, the turbo-compound engine passed a 150-hour navy qualification test in January 1950. The engine entered production, and the first unit was shipped in March 1950. Using

115/145 PN fuel, the R-3350 turbo-compound D-series was rated at 3,250 hp at 2,900 rpm for takeoff and 2,600 hp at 2,600 rpm for normal operation. The engine had a 56.59-inch diameter and weighed 3,520 pounds.

Initially, the turbo-compound engine was only available to the military, but in January 1952, the engine was scheduled for mass production for commercial aircraft. Although the turbo-compound was sponsored by the military, more than 20 percent of the engine's market was eventually commercial aviation.

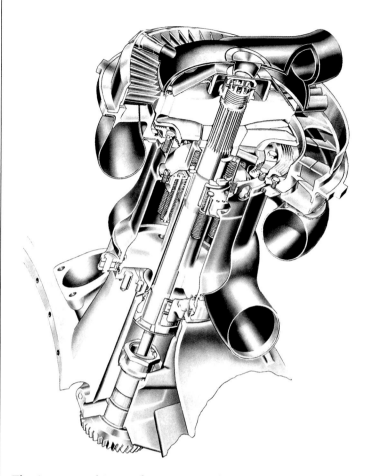

The inner workings of a PRT reveal another layer of complexity added to the piston engine. The exhaust gases from six cylinders spun the shaft-mounted turbine wheel. A gear at the bottom of the shaft engaged a fluid coupling (not shown). (Photo Courtesy Aircraft Engine Historical Society)

Development of the turbo-compound engine led to the R-3350 E-series, first produced around August 1953 and made commercially available in October 1955. The turbo-compound E-series had a takeoff rating of 3,700 hp at 2,900 rpm with anti-detonation injection and 3,400 hp without. The engine's normal rating was 2,800 hp at 2,600 rpm. The turbo-compound engine weighed around 3,650 pounds and had a diameter of 58 inches at its widest point.

Most turbo-compound engines had a 0.4375-propeller gear reduction ratio that was also fairly common on earlier R-3350 engines. However, the Lockheed L-1649 Starliners used R-3350-988TC18EA-2 (EA-2) engines that had a lower reduction of 0.355. Only 283 of these engines were produced for the 44 Starliners made. A few of these slow gear-reduction cases exist today, and they are coveted among air racers using the R-3350 engine. An EA-2 nose case attached to a specially

The PRT fed power back to the engine's crankshaft via a fluid coupling. Wright's research indicated PRTs provided more power for their weight than turbochargers, and the PRT did not require any additional instrumentation. (Photo Courtesy William Pearce)

The bevel gear of the PRT engaged another bevel gear that was mounted on a shaft that ran through the center of a spur gear and into the fluid coupling. The fluid coupling transferred power from the PRT's input to the hollow output shaft and its spur gear. The spur gear engaged a much larger gear coupled to the rear of the crankshaft and delivered power back to the engine that the PRT had harnessed from the exhaust. (Photo Courtesy William Pearce)

Wright R-3350 Turbo-Compound Engines in Airliners				
Manufacturer	*A/C Model*	*Engine Model*	*HP t/o*	*Number*
Lockheed	L-1049B, C, D, E, F	972TC18DA1	3,250	4
	L-1049B, C, D, G, H	972TC18DA4	3,250	4
	L-1049C, D, G, H	988TC18EA3	3,400	4
		988TC18EA6	3,400	4
	L-1649A	988TC18EA2	3,400	4
Douglas	DC-7	972TC18DA2	3,250	4
	DC-7B	972TC18DA4	3,250	4
	DC-7C	988TC18EA1	3,400	4
		988TC18EA4	3,400	4

Major Applications of the Wright R-3350 Turbo-Cyclone Engine	
Canadair	CP-107 (CL-28) Argus maritime patrol aircraft
Douglas	DC-7 airliner
Fairchild	C-119 Flying Boxcar transport (later versions)
	AC-119 G/K Shadow/Stinger gunship
Lockheed	L-1049C, D, E, G, H Super Constellation airliner
	C-121 C and later versions Constellation transport
	EC-121 Warning Star early warning aircraft
	R7V-1 Constellation transport
	L-1649 Starliner airliner
	P-2D Neptune maritime patrol aircraft
Martin	P5M Marlin flying boat

This ad from the early in 1950s extolls the virtues of the turbo-compound engine. Wright did not pursue jet engines with the same intensity as other manufacturers, and the company fell behind into a position from which it never recovered. (Photo Courtesy William Pearce Collection)

built R-3350 engine propelled the highly modified Grumman F8F Bearcat *Rare Bear* (race 77) to a world speed record of 528.329 mph on 21 August 1989. Successful Hawker Sea Fury racers September Fury (race 232) and Critical Mass (race 10) used R-3350 engines with EA-2 nose cases. Gear reduction for all R-3350 engines was through the use of planetary gears.

Not only did the turbo-compound produce more power than the basic R-3350 engine, it was also around 12 percent more fuel efficient.

In economical cruise at 1,400 hp, the turbo-compound had a specific fuel consumption of 0.378 pounds of fuel per horsepower per hour, which is the best value achieved by any production piston aircraft engine. However, hard-run turbo-compound engines did suffer some reliability issues.

The R-3350's appetite for exhaust valves resurfaced, and after being pulverized by the piston, the valve (and bits of piston) was ingested by the PRT, either severely damaging or destroying the unit. This led some to remark that PRT was a "Parts Recovery Turbine." Frequent engine shutdowns inspired a joke that the P&W R-2800-powered DC-6 was a four-engine airplane with three-blade propellers and that an R-3350-powered DC-7 was a three-engine airplane with four-blade propellers. Super Constellations were mockingly called the world's best trimotors.

The Wright R-3350 Turbo-Cyclone 18s in airliner service achieved a time between overhaul (TBO) of 3,000 hours, a great improvement over the 100-hour TBO of wartime R-3350s. The turbo-compound engine enabled nonstop flights in both directions across the United States and across the Atlantic. Turbo-Compound engines required close attention and careful operation. Reliability diminished rapidly if the engine was overstressed or operated carelessly. Production of the turbo-compound continued until 1961. Of the 50,185 R-3350 engines built, 11,954 (23.8 percent) were turbo-compounds. Wright was the sole producer of the turbo-compound engines.

Wright Versus Pratt & Whitney

Which company made better engines, Wright or P&W? This question inevitably comes up during an in-depth discussion on air-cooled radial engines. The honest answer is that both companies made incredible engines at the pinnacle of contemporary technology and power. The companies' products had many similarities and a few fundamental differences.

Some facts indicate that P&W was the superior company. By 1929, 90 percent of US commercial transports were powered by P&W engines. P&W supplied 50 percent of the power for US aircraft during World War II; Wright supplied 35 percent. P&W successfully transitioned to jet engines and is still in business today, whereas Curtiss-Wright, the second-largest US company at the end of World War II, exited the engine business in the 1960s and currently makes actuators, valves, and pumps. In the end, it was poor management, not inferior engines, that impeded Wright Aeronautical.

As previously stated, P&W was formed by Wright employees who left the company because management was not willing to provide resources to develop more powerful engines. This stance of Wright management did not change much as history moved forward. When P&W experienced bearing failures in the 1930s, nearly all of the company's resources were dedicated to solving the issue, one on which bearing suppliers had given up. After an incredible investment of time and money, the solution was found, implemented, and then shared with competitors because of wartime priorities.

Wright, on the other hand, had to persuade Curtiss-Wright to finance the development of new engines and to fund research to solve emerging problems. The company spent only what was necessary to get engines to an acceptable level. This had dire consequences when the R-2600 and the R-3350 engines were pushed to become more than test engines and actually entered production. Solutions to problems came slowly due to a lack of available engineers and resources. Wright management also had a tendency to blame other parties rather than seeking a solution.

With the R-3350, Wright felt engine cooling issues were a result of the aircraft manufacturer designing a poor cowling, and engine fires were the result of improper settings from the carburetor manufacturer. Resolutions came so slowly that NACA was enlisted to solve some of the R-3350's issues that Wright blamed on others. Wright slowly reacted to developmental problem, whereas P&W tried to anticipate the problems. This practice may have given Wright a more rapid initial development time and lower initial costs, but it cost the company time and money in the long run.

P&W took pride in subcontracting as much work as possible to other companies. All manner of parts were made by other companies and shipped to P&W to be assembled into aircraft engines. P&W relished the fact that it could get three competitive bids for something as mundane as cotter pins. Other companies that were industry leaders in their own specialized fields could build specific parts and send them to P&W for less than what it would cost P&W to make the parts itself. The active engagement of subcontractors also helped during economic downturns, as the financial impact was spread and absorbed among P&W and hundreds of suppliers, rather than P&W having to shoulder the burden alone.

When expansion was needed during World War II, P&W requested more from its existing suppliers and contracted new suppliers, enabling P&W to reach its production goals faster. P&W was happy to license its engines to automotive manufacturers during World War II, initially charging a $1-per-engine fee but later charged no fee. Ford was tasked to build the then-state-of-the-art R-2800 B-series, as P&W was not trying to hold on to its latest technological developments. During the war, approximately 63 percent of P&W engines were built under license; only around 29 percent of Wright engines were built under license.

Wright was a company forced to operate as lean as possible, and issues were encountered when vast expansion was needed. Wright wanted to produce its own engines, such as the R-2600. Studebaker was to build the R-2600, but Wright objected and forced a switch to the R-1820. Wright built all the R-2600 engines and would have probably built all the R-3350 engines if the demand had not been so high and if the government had not intervened. As additional Wright plants were built, quality suffered because management and truly skilled employees were spread increasingly thin. Some issues resulted from inadequate management, and other issues were a result of inadequate training. Regardless, the reputation of Wright engines was tarnished during the war, at least temporarily.

Wright and P&W both encountered major issues with production engines. P&W was able to solve the issues faster than Wright

Wright was understandably proud that its heritage dated back to the Wright brothers. This ad provides an interesting illustration of engine cylinder development over the years. Wright typically used larger cylinders than P&W. (Photo Courtesy William Pearce Collection)

P&W usually took a more conservative approach than Wright with its engines and its business. P&W's conservatism resulted in engines that combined power with dependability. No aircraft engine has matched the production numbers of the R-1830; this is one of the 173,618 produced in a test cell. (Photo Courtesy LOC 2017694144)

because P&W was not burdened by as many layers of management as Wright. P&W could begin work on an issue and did not need to wait for corporate bankers to allocate funds. P&W's employees built aircraft engines; Wright's employees worked for a company that built aircraft engines. This is not meant as a slight to the many Wright employees who took pride in their jobs. Rather, it is to point out the fundamental difference between how Wright and P&W viewed the manufacturing of their engines.

Wright and P&W's products were on par with each other. No classic aircraft or warbird operator is removing Wright engines from aircraft and replacing them with P&W engines. Wright and P&W engines built during wartime at the incredible rate of one every 10 minutes display the utmost quality and precision and are still running more than 70 years later. It was how the individual companies were run, not product quality, that determined the path for one company's long-term success and the other's failure.

Engine Mounts

In the early days of aviation, radial engines were attached to the airframe via a series of mounting pads positioned around the rear of the engine. As the supercharger and the accessory section grew, the mounting point remained relatively unchanged on the back of the crankcase, forward of the supercharger and accessory section. However, as engines became larger and heavier, this practice could not be maintained.

The Wright R-1820 G-series, R-2600, and non-turbo-compound R-3350s used a slightly different style of mount than what previous engines used. The mounting pads were positioned around the engine on the supercharger housing, between the intake pipes and as close to the cylinders as feasible. This position was selected to locate the pads as close to the engine's center of gravity as possible. Due to the tight fit, the mounting pads were parallel to the engine and not perpendicular, as was the normal practice. The engine mount extended from the mounting pad and to the engine mounting ring. The engine mounts around the mounting ring were angled (conically disposed) so that if an imaginary line extended from the mounts and through the engine, the lines from all the mounts would converge at the engine and propeller's center of gravity. These imaginary lines were also tangent to the mounting ring. This mounting technique dynamically suspended the engine and isolated vibration about the roll, pitch, and yaw axes. In addition to drastically lessening engine vibrations, the mounts provided better support.

Wright's "dynamic suspension" mounting design was developed by Kenneth Browne of Wright Aeronautical and Edward Taylor of the Massachusetts Institute of Technology (MIT). However, some elements of the design infringed upon patents previously awarded to Hugh and Thomas Lord of the Lord Corporation. The two companies were able to resolve their differences and entered into an agreement in 1939. Because Wright made engines, and the Lord Corporation made mounts, the Lord Corporation purchased the patent from Wright to exclusively produce the dynamically suspended engine mounts. In return, Wright agreed to purchase the mounts exclusively through the Lord Corporation. A short time later, the Lord Corporation named the mounting system the Dynafocal engine mount.

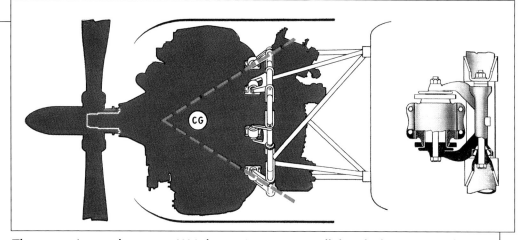

The mounting pad on most Wright engines was parallel with the crankshaft. The engine mount extended from the mounting pad to the mounting ring at an angle. An extended line through all mounts converged roughly at the engine and propeller's center of gravity. (Photo Courtesy **The Wright Cyclones,** *1942)*

At the same time, P&W was developing the R-2800 and approached the Lord Corporation to design a new engine mount for the large, heavy, and powerful engine. P&W's design differed from Wright's in that the entire mounting pad was angled. However, P&W's design still used the concept of lines extending from the mounting pads and

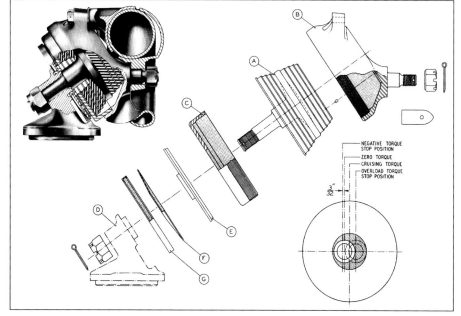

This image shows a breakdown of the standard Dynafocal engine mount used on P&W engines. The core assembly, made up of alternating layers of rubber bonded to steel plates (A) was held in a housing, by covers (B), a bracket (C), and was attached to the engine's supercharger housing (D). A friction damper (E) was loaded by a dished spring (F) and a retaining plate (G) when the mounting nut was tightened. (Photo Courtesy **Installation Handbook,** *March 1948)*

intersecting at the engine and propeller's center of gravity. The style of engine mounting was a further development of the Wright design but was still considered a Dynafocal mount. Very quickly, the Lord Corporation became the sole supplier of engine mounts for all the large radials built in the United States. The company experienced a 64-fold increase in employment, growing from 70 employees in 1939 to a peak of 4,500 employees during the war.

The angled Dynafocal mounting pads were used on the R-2800, R-4360, and Wright R-3350 Turbo-Compound. The pads were still positioned on the supercharger housing, as close to the cylinders as possible. In the R-3350 Turbo-Compound, the mounts were on the backside of the PRT housing. For the R-4360, a bracing bar extended from each motor mount and attached to the crankcase between the A and B cylinder rows on A- and B-series engines, and to the B and C cylinder rows on C-series engines.

With a few exceptions, the motor mount was where the engine designer's responsibility ended. The aircraft designer was responsible for mounting the engine to the airframe. A good deal of engineering was involved to create engine supports that withstood the stresses of a 3,500-pound engine with a whirling, 800-pound, 17-foot propeller, both subject to constant vibrations and occasionally loads in excess of several times the force of gravity.

My Life with Connie

By Judy Whiting James

In January 1955, a young woman from Boston, Massachusetts, answered a large newspaper ad for a "cattle call" for stewardesses (as they were called in the old days) in the hopes of being hired for the only job she ever wanted since the age of 12. The interview with Eastern Airlines went very well at the Parker House Hotel in the Back Bay area of Boston, an up-scale locale in its day and still today. Facing a panel of four male interviewers was a bit daunting.

An unusual request came at the end of the interview when I was asked to raise my skirt about three inches above my knees because dresses were a bit longer in those days. I found out later at the training school that it was just a test to see how flustered you would become in a precarious setting. (You can believe that if you want to!)

After two in-person interviews and one telephone interview, I was hired and left home two weeks later for Miami on a snowy Valentine's Day, going from freezing to roasting in a bit more than six hours. Onboard the flight were two other girls, and we all became good friends and roommates. We were housed near the airport and training center. After four weeks of training at Miami Airport (MIA), then located on W. 39th Street, in a classroom in the maintenance hangar led by a very enthusiastic "Music Man" type of instructor, we were ready to be launched into another realm. There were 15 in the class and beginning seniority was determined by age. As I was the youngest in the class, I was instantly the "baby" of the base, so I never thought I would get my bid to remain stationed there. I was thrilled when I got MIA, my first choice. It seems everyone just had to get New York, and those who did had a very difficult time finding lodging in Manhattan.

Six girls to a tiny apartment was not unusual. Thankfully, all six were never in residence at the same time due to trip assignments. I never got enough seniority built up to hold a "bid" trip, as no one wanted to transfer out to another base. So I flew reserve for the usual two years that the majority of girls flew before leaving to marry, as back then married stewardesses were not allowed. Every day was a new adventure, as I had no idea where I was going the next time out on the fast-expanding route system. Chicago was the farthest west Eastern went in those days, to Midway airport, as there was no O'Hare even on the radar at the time.

My very first working trip was to Chicago and a quick turn-around trip besides, so that was an unforgettable and exhausting day and also my "check ride." The inflight service supervisor was onboard checking my every move. I am told my first safety announcement sounded as if I meant for everyone to jump out right then and there! Fortunately, just like today, nobody was really paying any attention anyway. Because I never gave crew-scheduling a hard time, nor complained about where I was going, when Boston trips needed filling, I was the one they called, and I could sleep in my own bed that night at home after walking from Miami to Boston. It was literally the best of both worlds.

My two new (older) friends and I rented a little house in Miami Springs within walking distance of the terminal and fell asleep at night to the incoming engine noise of flights from all over flying right over our house. To this day, piston engines are music to my ears.

We did not know for a while that our landlord was a member of the horse racing world's mafia. He was such a very nice gentleman. I guess we three "proper Bostonians" still had a lot to learn. Our rent was $100 a month and even each of our shares was a struggle to save, at $33.33 each. We took turns throwing in the extra penny. No one would guess how many starving stewardesses there were serving delicious meals to the passengers. Hopefully the airline never figured out where all those untouched cherry tarts went!

Because luggage in 1955 was still very heavy and with no wheels and just a handle, we walked from the house to the terminal and then downstairs to Operations for check-in at one hour prior to crew boarding the airplane for preflight preparations. What a treat it was to be offered a lift out to the plane from a baggage handler (truly a knight in shining armor on a tractor). Hold on to your hat! The Constellations being so large were not parked close-in to the terminal.

Prior to leaving Operations, we were checked for weight (ours) and all that went into a proper-looking stewardess, even an occasional girdle check (or I should say "snap"). That certainly wouldn't "fly" in these days! Our pre-flight duties consisted of folding cocktail napkins and/or dinner napkins just so, checking to be sure that enough meals were onboard by the caterers, a cabin walk-through checking for enough blankets, pillows, magazines, and fresh seatback covers. There were no pull-down tray tables in those days or covered overhead racks. Carry-on bags were stowed under the seat in front of the passenger.

The plane cleaning crews did a fine job making the cabin ship shape (make that "plane shape"). We also had to make out seat checks that were affixed to the seatbacks of each seat. They were for the purpose of identifying hung-up clothing to be reconnected with the correct passenger at his or her destination and also for taking special requests, particularly at meal time. We used a lot of black grease pencils! What a pleasure to have wide aisles, large windows at each row, two seats per row, wide comfy seats and larger lavatories and galleys.

At boarding time, often a ramp agent stood at the bottom of the stairs, but most of the time passengers were welcomed by a stewardess, depending of the type of aircraft being flown and the number of cabin crewmembers needed. The stairs used for the Constellations were quite steep, and at the top was a stewardess on the platform surveying her kingdom and greeting her subjects like the queen of the universe. I knew no greater thrill!

After boarding all the passengers and getting an accurate walk-through head count that had to match exactly with the manifest, our sleeping giant then became awake, and the ramp agent ran out to the side window just to the left of the captain's seat where the chart with the total weight of cargo, people, fuel, and all else was handed up on the end of a long pole that resembled an apple picker.

Meanwhile back in the cabin, the cabin crew settled passengers and checked for seat-belt compliance, offered magazines, answered questions, hung coats, and stowed into the closets any too-large items that did not fit safely under the seat in front. Concurrent to our bustling inside, the luggage outside was loaded beneath us. That noise alone was unnerving to some flyers who thought we were already doomed and they had to be reassured as to what was really happening below. The flight engineer did a walk-about outside and underneath checking visually for any anomalies that could be attended to before taking off.

If some of this is a bit out of order, it was because I was too busy inside to pay much attention as to who did what and when. I was just thankful that it all got done by very competent and skilled employees. I loved the whole procedure so much that each time I boarded an empty airplane I was engulfed by the heady aroma of aircraft fuel, upholstery, galley odors, and cleaning fluids; just all the best of the aviation world's perfume.

After all the passengers were boarded, assorted smaller details needed to be attended to, which could be done while taxiing out to the runway to await our turn to take off into the blue yonder. Last-minute announcements were mandatory before the cabin crew rushed to their own seats prior to the rush down the runway (all part of the excitement of the job).

Another sound that unnerved new flyers was the loud thumping that the landing gear made from beneath the seating areas both at takeoff and again at landing. I loved to perform all this calming helpfulness so much because it gave me much satisfaction to put any doubts and fears to rest. At the end of a flight for a fearful flyer all I wished for was that I had made a person at ease and ready to fly again.

After receiving "the word" from the pilot, it was time to begin our in-flight duties. No two trips were exactly alike. Although consistent by regulations, they were different because the people themselves were different. Because there was no First Class or Coach Class, I liked to believe that all of our passengers were considered First Class. We certainly tried to make it so. One way to succeed in that respect was to make any request seem to be absolutely no trouble at all.

On a perfect flying day, two stewardesses began the beverage service prior to serving lunch or dinner. A third stewardess worked in the galley and was indeed as busy as the others. Alcoholic beverages cost $1.00 and soft drinks and "coffee, tea, or bullion" were no charge. At the end of a flight back at home base, the senior stewardess tallied the empty miniature bottles with the money collected and prayed it all came out evenly. Sometimes, because of making incorrect change, it did not, and out of fatigue we threw in our own money just to get home to sleep! But I get ahead of myself.

The meal service was usually quite smooth and orderly when the flying conditions were just right. If there was much turbulence before all of the procedures began, we had to delay in-flight services until further instructions from the captain. Not a fun thing at times when the rear of the cabin was still eating dinner as we were

A typical day in the Northeast, one like any other for the cabin of Eastern's Lockheed Super Constellations.

beginning to collect food trays from the front. Prior to any beverage or food service we distributed travel-size pillows on which to place the food trays as we could only carry two trays at a time. On a full flight that would be items to 88 passengers plus the crew; that equated quite a lot of trips back and forth to the galley.

Most of the time there were three choices of main dish: beef, chicken, or fish. If turbulence occurred during the meal service, we discontinued it until further notice from up front. There were instances of just barely finishing a meal service before beginning our final destination preparations. A big secret: There were times when I slid into my seat just in time for the airplane to touch down while hanging on to my seat-belt buckles hoping no one would notice!

Next, it was time to help the mothers with infants and toddlers to deplane. Actually, it was a fun thing to witness reunions among the friends and families. The moms were advised to carry the infants down the stairs and the stewardess to walk the toddlers very carefully down the stairs and carry the teddy bears etc. as well.

Time for the cleaning crews to hop into action to prepare for the next day's trip and leave everything fresh and clean again. They were the unsung assets to making a long trip very pleasant.

At the end of a six-hour "walk" from New York to Miami it was a thrill to glide past the big red "Fly Eastern" sign at the edge of the runway. The happy yet sad finale to a busy day and already I was wondering where I would be going the next day. (Needless to say, to this day I am a very fast suitcase packer and ready to answer the call to go flying!)

All of the airplane noises and odors and hubbub were like a great symphony to me; invigorating yet soothing at the same time, perfect testimony that the world truly is turning.

A bevy of lovely ladies of the sky, these Eastern stewardesses are wearing their summer uniforms and standing before a 749 model Constellation. Note the Speed Pak attached.

Vintage photo of Judy Whiting James (far right) underneath the tail of one of the airline's 1049C Constellations. Sixty-plus years later, I can tell you she hasn't changed much at all. (Photo Courtesy Judy James)

TWILIGHT OF THE GODDESSES

As the decade of the 1950s hit midstride, commercial aviation approached an apex relative to airplane powerplant technology, with both of the fierce competitors in Southern California vying for the ultimate championship in design prowess. One more growth model was left inside the Douglas and Lockheed propliners, and each was about to enter service with international airlines. Unacknowledged by the companies, however, this heralded the end of the great era of piston engine air travel. Much like a supernova that burns brightest and then extinguishes itself, the time period for the "world's ultimate propliner" was white-hot but noticeably brief.

Perhaps the greatest photo in this book, and looking as though it was shot yesterday, this is the extremely rare TAI Douglas DC-7C on a pre-delivery flight. The image is rare because this was the only airline to fly the airplane type without the trademark radar nose. (Photo Courtesy Boeing Co. via Jon Proctor with permission)

Kodera

Getting There by Air

In 1956

The last remaining hurdle for the postwar airliner to surmount was that of range. This recurrent theme was about to turn a corner with the introduction of the DC-7C Seven Seas and the 1649A Starliner Constellation. Please bear in mind that it was then a mere 10 years after our greatest war and 15 years after the Boeing 314 when suddenly, everyday people could fly nonstop from Los Angeles to Paris or London and return in similar fashion, with no worries of weather or winds. This was *huge!* (And just another short 10 years later, jetliners flew even farther yet, such was this exponential increase in capabilities in airline flying.) The key was to enlarge the airplanes so they could carry the requisite additional fuel tankage to master this feat.

The Final DC Propliner

PAA approached Douglas once more and asked for just such a long-range version of the DC-7. Thanks to another uprating of the R-3350TC engine, PAA's requested airplane was viable. The resultant airplane would be 42 inches longer to allow for more passengers and baggage/cargo, have fuel capacity increased by grafting in stub wings between the existing wing and fuselage, enlarge the vertical stabilizer due to higher-powered engines, continue the use of the saddle tanks behind the engine cowling areas in the nacelles, and enhance systems and furnishings so as to be compatible with overwater flight.

A great benefit of widening the wingspan was to increase the distance from the passenger cabin of the Wright Turbo Compound

This glamour shot shows the prototype DC-7C in its red, white, and blue scheme. Notice that the nacelle-saddle fuel tanks are painted white and stand out dramatically. The additional fuselage stretch is obvious, along with the taller vertical stabilizer and radar nose. (Photo Courtesy Museum of Flying Collection)

Taking a trip to Europe? Fly PAA. Please observe the dress code and enjoy approaching the aircraft along the Pan American Blue carpet.

engines, thus lowering the noise within the interior. Nice touch. Also, the greater wingspan created a higher aspect ratio to the wing, which lowered the induced drag, which helped with speed and range. With all these embellishments, the Seven Seas looked like

This is a rare photo of even-more-rare markings for SAS: bare metal. This look was a hybrid of the then-current DC-6 scheme and did not last very long at all. The modified "jet" design with white crown fuselage came along soon thereafter as the Sud Caravelle entered service with the airline.

quite the thoroughbred, a fine improvement over the basic DC-7.

A total of 13 airlines ponied-up to buy the airplane, perhaps most interesting of all, British Overseas Airways Corporation (BOAC), the British flag carrier. Its new Comet 4 was three years in the future as was the Britannia 312 series turboprop, and the Stratocruiser was a bit too large and slower than the DC-7, so as an interim solution BOAC bought from Douglas.

The first flight of the DC-7C was in December 1955, and by May 1956, PAA had the airplane in service from New York to London. A total of 121 airplanes were delivered until Swissair took the final airplane in November 1958.

With the introduction of the DC-7C, airlines discovered the enhanced value of routings flown over the North Pole. SAS had pioneered such flying with their DC-6s a few years earlier, and now PAA, KLM, and Japan Airlines took advantage of the airplane and its range by following suit. Typical eastbound flight times to London were 18 to 19 hours.

As a final bow to the heated competition with Lockheed, the Douglas airliner was in service a full year earlier than the rival Starliner, leaving TWA to play catch-up with its own over-the-pole flights from Los Angeles to Paris. The Starliner may have had the last laugh, however.

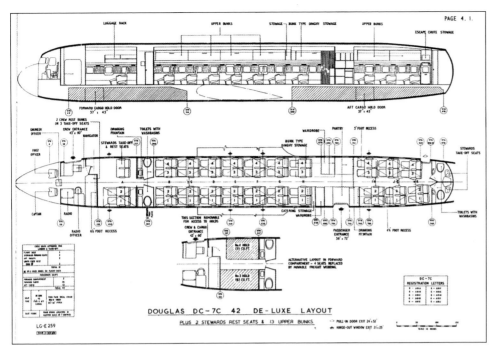

LG-E 259

DOUGLAS DC-7C 42 DE-LUXE LAYOUT
PLUS 2 STEWARDS REST SEATS & 13 UPPER BUNKS

De Luxe seating layout of a Seven Seas: 42 passengers accommodated in sleeperette-style seating for the long overwater trips of this airplane. Note also the rather ignominious location of the steward takeoff seats: in the lavatories!

Lockheed's Star Performer

Spying on what Douglas and PAA were concocting on the west side of Los Angeles, TWA dropped by Burbank to cajole Lockheed into an appropriate rejoinder to the DC-7C threat. Work started in 1955 on a long-legged version of the Constellation, and the result was an interesting combination of old and new. The 1649 used the same fuselage as the 1049G series but had an entirely new and broader wing, engine nacelles, propellers, and landing gear. As with the DC-7C, the Starliner's wider wingspan reduced the induced drag yielding the predicted positive range results.

Tip tanks were once again offered for the Starliner. The wing now measured 25 feet wider, totaling 150 feet in all. The airplane started to look like a sailplane, and the impressiveness of this airfoil was exaggerated and highlighted by Howard Hughes' favorite illustrator Ren Wicks in the original advertisement for TWA's newest fleet member. This final version of the Connie was heralded as the queen of the skies.

Combined with the same distance-noise-reducing concept found on the DC-7C, more insulation was padded into the fuselage, and the propellers geared down to run quieter.

It is now apparent just how peaceful and efficient these two finalists stood as they approached the finish line. Propeller aviation ended with great aplomb and a lovely flourish of pomp and luxury.

The Connie flew faster than the Seven Seas and

The prototype of the Jetstream for TWA, N7301C, was identified by the extended pitot/test boom on the right wingtip. Note the dihedral and massive wingspan of this airplane. (Photo Courtesy Jon Proctor)

The first flight of the prototype 1649 as it leaves Burbank's runway heading south. Obvious here are both the increased wingspan as well as the very wide track of the main landing gear. (Photo Courtesy Jon Proctor)

Wings span the globe for Air France. Art by Richard Dean Taylor illustrates just how far the 1649 could roam, which in the case of the French carrier, was quite a long distance indeed.

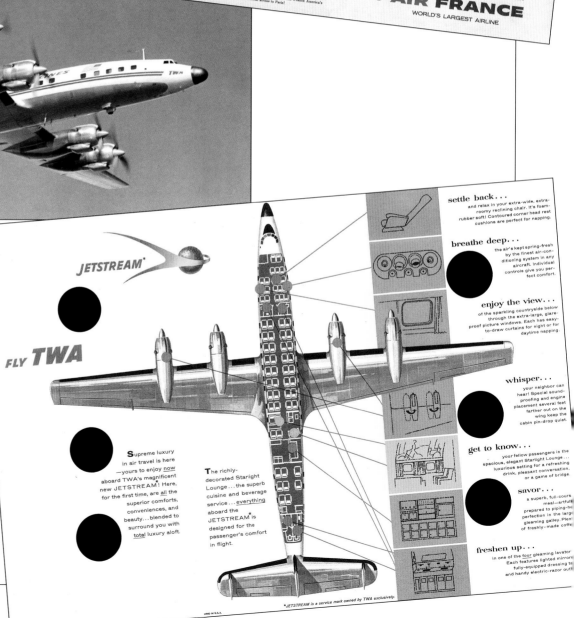

177,000 mile wingspread

This is Air France! Larger by far than any other air network in the world. Stretching eastward from the interior of South America to the vast reaches of the Pacific. Connecting 215 cities in 76 countries. Spanning 5 continents. Telescoping distance into hours. Flying the Jet Stream over the Atlantic in new Lockheed Super Starliners—world's fastest longest-range, most luxurious airliners. And on every flight, whether it be from Brazzaville to Dakar, New York to Rome, Chicago to Paris, Boston to Tokyo, you find the same superb service, the same high standards of dependability that have made Air France America's favorite airline to Paris!

✦ **AIR FRANCE**
WORLD'S LARGEST AIRLINE

Inflight view of the first flight of N7301C. The ultra-wide wing seemed to stretch from horizon to horizon. (Photo Courtesy Gerald Balzer Collection)

Seating layout and amenities offered in the Jetstream by TWA. Thinking in terms of propliner aviation, the First Class section with the Siesta Seats was located, as always, in the tail section of the airplane, allowing for eight passengers to be carried in ultimate comfort. Conversely, the forward section looks almost like any Coach section on today's airplanes. Many times these were crew rest seats.

JETSTREAM*

FLY TWA

settle back . . .
and relax in your extra-wide, extra-roomy reclining chair. It's foam-rubber soft! Contoured corner head rest cushions are perfect for napping.

breathe deep . . .
the air's kept spring-fresh by the finest air-conditioning system in any aircraft. Individual controls give you perfect comfort.

enjoy the view . . .
of the sparkling countryside below through the extra-large, glare-proof picture windows. Each has easy-to-draw curtains for night or for daytime napping.

whisper . . .
your neighbor can hear! Special sound-proofing and engine placement several feet farther out on the wing keep the cabin pin-drop quiet.

get to know . . .
your fellow passengers in the spacious, elegant Starlight Lounge . . . luxurious setting for a refreshing drink, pleasant conversation, or a game of bridge.

savor . . .
a superb, full-course meal—artfully prepared to piping-hot perfection in the large gleaming galley. Plenty of freshly-made coffee

freshen up . . .
in one of the four gleaming lavatories. Each features lighted mirrors, fully-equipped dressing table and handy electric-razor outlet.

Supreme luxury in air travel is here —yours to enjoy _now_ aboard TWA's magnificent new JETSTREAM! Here, for the first time, are _all_ the superior comforts, conveniences, and beauty . . . blended to surround you with _total_ luxury aloft.

The richly-decorated Starlight Lounge . . . the superb cuisine and beverage service . . . _everything_ aboard the JETSTREAM* is designed for the passenger's comfort in flight.

JETSTREAM is a service mark owned by TWA exclusively.

This staged photo from TWA shows the food service and wide standard seats found onboard a typical 1649 flight. Notice the Lockheed penchant of not using trays in seat arm mounting but rather the old style of eating food perched atop pillows.

boasted that all the capitals of Europe were now within unrefueled flying distance of New York. Both TWA and Air France used their aircraft over polar routes to the United States. Because the emphasis was on long range, payload was not as important as being able to get there. The Lockheed airplane typically carried 58 or maybe 64 passengers in a four-abreast, all–First Class configuration. Berths were not a part of this Connie's interior furnishings, but again the sleeper seat made its appearance throughout the cabin.

A 17-hour flight to Europe was made quite comfortable. All airlines could order the cocktail lounge found in the Super G model, and TWA signed up for it instantly. Once again, Henry Dreyfuss designed wall coverings to enhance the ambience of Starliner travel. As has been true with many other airplanes discussed in this book, international travel by 1649 was again an acknowledged "true experience" but was now approaching flawless and totally civilized. Airline flying had come a galaxy's worth of distance from the Pitcairn and Ford days of the 1920s.

Au Revoir

So here we are, at the end of a frenetic, pitched battle for air supremacy among the nation's top two airliner constructors. By 1958, the last of the Lockheeds had been delivered as had the end-piece from Douglas. No other nation on Earth could have produced

such sterling engineering ensembles, giving wings to a tired and beaten world. The competence of design and designer was honed to a diamond-edge fineness. All the technical leadership endowed to these two companies stood the world of aeronautics in good stead indeed for many decades following the great propliner slug match.

The zenith of radial-engine passenger airplane development, evidenced in two different airplanes, was all too short lived, almost to the point of cruel joke. All of the major jetliners were on scene plying the world airways by 1960, a scant four years after the introduction of the big props. The queens were given the broom and relegated to converted cargo-carrying duty by the major airlines. And even that didn't last terribly long as jet freighters rolled off the lines at Boeing and Douglas. The shadows grew long; the flying public was now enamored with the wonderous jet and its promise to mankind.

All things were now going to change, and so, we carefully wrapped and packed our warm and sultry memories in beautiful rose-colored tissue and crepe, shuttering away the old days for another time and another generation. The world had crossed yet one more Rubicon, and the most amazing era in passenger transportation had drawn to its inevitable and irrevocable conclusion.

Pearce

Hardware on the Wing

Piston Engine Postscript

After World War II, Wright and P&W became increasingly focused on developing jet engines. During the war, Wright had developed the R-4090, with 11 R-3350-cylinders per row, to compete with P&W's R-4360. However, Wright stopped work on its 22-cylinder engine as problems arose with the R-2600 and R-3350. The R-4090 was the last new engine developed by Wright.

In 1946, P&W developed the R-2180E. This 18-cylinder engine incorporated everything P&W had learned over 20 years of developing air-cooled radials. The R-2180E used two rows of cylinders developed from the R-4360, a built-up crankshaft, and single-piece master rods. Despite P&W offering it as a replacement for the R-2000s installed in the DC-4, the R-2180E engine did not find a place in the US market, and only 60 engines were built. It was clear to everyone that jet engines were the future of aviation.

During World War II, the US government prevented aircraft engine companies from developing jet engines; instead, the companies focused on improving existing piston engine technology. Companies that had experience with steam turbines were given US government support to develop jet engines, which is why General

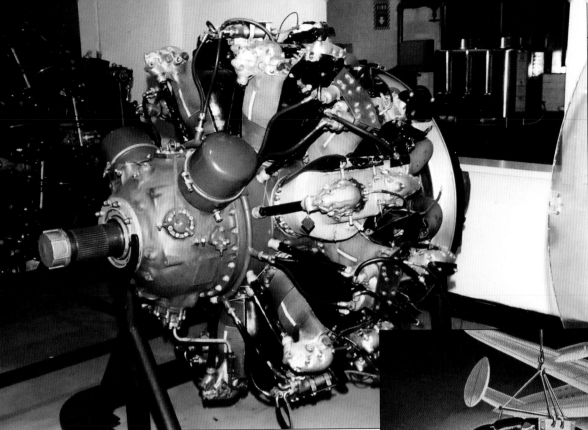

Electric and Westinghouse entered the aircraft engine field: These companies had experience with steam turbines for power generation.

Wright, P&W, and other aircraft manufacturers had a lot of catching up to do to develop a first-rate jet engine. Both Wright and P&W were able to rise to the challenge, but Wright struggled and was less focused than P&W. Publicly, Wright continued to espouse the piston engine into the mid-1950s; after all, the company had just developed its turbo-compound engine. Wright's jet engine development program could not match that of P&W, and Wright was ultimately dragged down by other divisions of the Curtiss-Wright Corporation. Today, P&W is a world leader in jet engine development and production.

Hundreds, if not thousands, of aircraft continue to use Wright and P&W radials of all types. Although some parts for these engines are incredibly scarce, a few companies have

This ad from the 1940s depicts Wright's past, present, and what the company hoped for its future. Wright believed that its R-3350 would continue to power aircraft beyond the 1950s. Unfortunately for Wright, the engine was not powering any new aircraft as the 1960s dawned. (Photo Courtesy The Wright Cyclones, 1942)

risen to the challenge and have begun manufacturing new parts. Without a doubt, Wright and P&W engines will continue to fly for decades to come.

Large, piston engine transports continued to soldier on into the jet age, but their numbers were in a constant state of decline as jet engine transports took over. Although DC-3s continue to operate in limited capacities around the world, C-46s, DC-4s, and DC-6s have become nearly extinct, operating only in remote places such as Alaska or South America. Many round-engine airliners were used as firebombers, but even those have been almost completely replaced by turbine- or jet-powered aircraft. A few DC-6s, one DC-7, and a few Constellations have been restored and are operated on the airshow circuit. With luck and a lot of money, more transport aircraft will join them to ensure that propliners are remembered for many years to come.

Major US Radial Engines Used in Airliners											
Year	Manufac-turer	Engine	Cylin-ders	Bore (inches)	Stroke (inches)	Displacement (ci)	Initial Power	Developed Power	Diameter (inches)	Length (inches)	Weight (pounds)
1924	Wright	J-4 Whirlwind	9	4.5	5.5	787	**200 hp** at 2,000 rpm		44	34	475
1925	P&W	Wasp R-1340	9	5.75	5.75	1,344	**425 hp** at 1,900 rpm	600 hp at 2,250 rpm	51.75	43	650
1926	Wright	J-5 Whirlwind	9	4.5	5.5	787	220 hp at 2,000 rpm		44.25	34	510
1928	P&W	Hornet R-1690	9	6.125	6.375	1,691	**525 hp** at 1,900 rpm	875 hp at 2,300 rpm	54.5	45	750
1928	Wright	J-6 Whirlwind R-970	9	5.0	5.5	972	300 hp at 2,000 rpm	450 hp at 2,250 rpm	45.0	45	485
1929	P&W	Hornet B R-1860	9	6.25	6.75	1,864	**575 hp** at 1,800 rpm	650 hp at 2,000 rpm	57.0	45	985
1929	Lycoming	R-680	9	4.625	4.5	680	225 hp at 2,100 rpm	330 hp at 2,300 rpm	43.5	34	515
1931	P&W	R-1830 Twin Wasp	14	5.5	5.5	1,829	**750 hp** at 2,300 rpm	1,350 hp at 2,800 rpm	48.0	61	1,162
1932	Wright	R-1820 Cyclone 9	9	6.125	6.875	1,823	575 hp at 1,900 rpm	1,525 hp at 2,800 rpm	54.25	48	920
1936	P&W	R-2180 Twin Wasp	14	5.75	6.0	2,181	**1,400 hp** at 2,500 rpm		51.63	63	1,675
1936	Wright	R-2600 Twin Cyclone 14	14	6.125	6.3125	2,604	**1,500 hp** at 2,400 rpm	1,900 hp at 2,800 rpm	55.0	55	1,935
1937	Wright	R-3350 Duplex Cyclone 18	18	6.125	6.3125	3,348	**2,200 hp** at 2,800 rpm	2,700 hp at 2,900 rpm	55.78	76	2,668
1937	P&W	R-2800 Double Wasp	18	5.75	6.0	2,804	1,800 hp at 2,600 rpm	2,800 hp at 2,800 rpm	52.8	80	2,360
1940	P&W	R-2000 Twin Wasp	14	5.75	5.5	1,999	1,300 hp at 2,700 rpm	1,450 hp at 2.700 rpm	49.1	60	1,570
1941	P&W	R-4360 Wasp Major	28	5.75	6.0	4,362	**2,800 hp** at 2,600 rpm	4,300 hp at 2,800 rpm	52.5	97	3,500
1947	Wright	R-3350 Turbo-Cyclone 18	18	6.125	6.3125	3,348	**3,250 hp** at 2,900 rpm	3,700 hp at 2,900 rpm	58.0	90	3,650

Ready for boarding at Los Angeles International Airport with promo attire quite prominent.

By Jon Proctor

TWA, a staunch supporter of Lockheed's development of the Connie in competition with the four-engine Douglas propliners, first began scheduled nonstop, eastbound segments to London and Paris with the Model 749A ("f" symbols in timetables indicated "fuel stop(s) may be necessary"; westbound trips were scheduled with an en route stop).

The 1049G Super G more reliably avoided Gander and Shannon visits to add avgas, but consistently nonstop crossings by TWA came about only when Lockheed's 1649A Starliner was introduced by the airline in June 1957 (Airways, June 2007). Its range permitted the first westbound nonstops to New York from such southern European cities as Rome.

The marketing name "Jetstream" was adopted for this last of the Constellation series, with Howard Hughes given credit for the idea, although others claimed to have thought it up. Of course, the name conjured visions of a jetliner, but the corporate excuse was that it related to the Starliner's ability to take advantage of jet-stream winds.

On 29 September, twice-weekly nonstop Polar service from Los Angeles to London began, flights that took more than 18 hours to complete. Celebrities on the inaugural run included actress Donna Reed. On 1 October, the first westbound trip, from London to San Francisco, found its way into aviation folklore, having endured the longest time aloft for a scheduled landplane segment.

Flight 801, fully loaded with nearly 10,000 gallons (37,850 liters) of high-octane fuel, wasted no time after leaving the gate at Heathrow Airport and lumbered off the ground to begin the 5,300-mile (8,530-km) trip. The augmented crew included Captain Gordon Granger, TWA's international division director of flying, along with three other pilots, two flight engineers, and two navigators. Purser Gerard Miston and two hostesses looked after 32 passengers, including tennis star Pancho Gonzales. Timetables listed the scheduled time aloft as 22 hours and 5 minutes.

The trip was subject to a fuel stop, if necessary, at Frobisher Bay (now known as Iqaluit, Nunavut), the principal of three approved alternates in Canada, where a TWA mechanic was permanently based. However, Granger was determined to fly nonstop to California. As a conservation measure, he filed a flight plan using super-long-range cruise procedures. The plan was to climb initially to 7,000 feet, but two hours after leaving London, over the Irish Sea, the Starliner had reached only 2,000 feet. At 4,000 feet, climb power was reduced, letting the airplane gradually ascend consistent with weight reduction as fuel was burned off. The planned initial altitude was finally attained as the airplane approached the southern tip of Greenland. An airspeed of 200 knots was maintained during most of the flight.

As TW801 passed abeam Winnipeg, the crew received a re-release to San Francisco from TWA Flight Dispatch and pressed on. Clearing the Rocky Mountains required a climb to 16,000 feet. Over Pocatello, Idaho, a gradual descent was begun, again conserving fuel as the airplane drifted down. Power was further reduced as the altitude bled off and, according to flight engineer Jerry Zerbone, "Our engines were barely turning over."

Following the Humboldt River into the San Francisco Bay Area, descent continued as the airport came into sight. With Oakland Airport visible on the left, just in case, Captain Granger completed the landing on Runway 1 and pulled into the gate 23 hours and 31 minutes after blocking out at Heathrow. The Starliner had been in the air for 23 hours and 19 minutes, a record that stands to this day.

Along with a Lockheed technical representative, an official from the CAA was on hand to check the amount of fuel remaining. Three hundred gallons (1,135 liters) were required to be "legal," which is exactly what was collectively left in the four main tanks; the other five were empty. Remarkably, the same cockpit crew reboarded flight 801 and flew it on to Los Angeles.

Operating the Jetstream Polar flights nonstop was more a matter of prestige than efficiency for TWA. The added weight of full fuel loads prevented cargo carriage and burned more avgas in the process. Using normal cruising speeds and making a stop would actually have resulted in less flying time.

Although "going direct" may have been popular with customers, these epic journeys did not last for long. TWA's introduction of trans-Atlantic jet service in November 1959 allowed much quicker connections between California and Europe via New York, and the Polar flights were discontinued. In May 1960, Boeing 707s began operating from Los Angeles to Paris with a fuel stop in Montréal, reducing the en route time to 12 hours and 30 minutes. The Canadian stop was eliminated when 707-331B Intercontinental StarStreams took over on 15 December 1962; 747s replaced the 707s on 31 October 1971.

The airplane from Jon's story is finally resting at San Francisco International Airport.

Many Typical Days

By Judy Whiting James

Many times through the years since my flying days an amusing thought still creeps into my head, probably triggered by some current happening. One night while eating some carrots and peas, I was reminded of gathering turbulence during a full-flight dinner service on a Super Constellation. The cabin crew was really hurrying to complete the serving of the main part of the meal when the mother of a capable-looking young boy asked rather haughtily if I would come back and separate his carrots and peas as he did not like different foods touching each other. As the turbulence was really increasing, I said I would be right back when all the other passengers were served but honestly cannot remember if I ever did return. That sort of comes under the same cover of all the aspirin I never delivered, probably because *I took them instead!*

On rare occasions, a notable person would plead with the powers that be to have the flight departure delayed a little, really quite a no-no but done on special request. Late one evening at LaGuardia (New York) for a flight to Miami, a well-known Broadway and movie musical star came rushing onboard in a heavy coat covering her very skimpy show costume. She disappeared quickly into the lavatory to change into comfortable clothes. She was in there quite a while when I heard a loud "PSSST!" and then saw a hand motioning me to come in. The long zipper in the back of her costume had become stuck halfway down so I had to try (successfully, as it turns out) to get it fixed and working. It could have been a long, chilly night for one very harried actress.

On a daytime flight about to depart Washington National, we were busily hanging up coats, checking seat belts, and answering assorted questions all the while taxiing out for takeoff. A sweet elderly woman kept trying to tell me the wing was on fire. I did not see anything but a burst of flame from the piston engines on the side and told her not to worry as that was quite normal (in those days) and would soon stop. Well, it did not stop and became quite a bit worse during the climb as we took off. I finally got a good look at it and went up front to tell the captain, and he came back to look at it himself. In a very few seconds we were headed back down, being greeted with heavy foam along both sides of the runway and flanked by fire trucks. The cause? A refueler had neglected to put the fuel caps on and fuel was burning off as it was sucked out during the ascent into the wild blue yonder. Soon we were on our way again, and the woman was rewarded with an extra cherry tart with her dinner!

One of my very first flights was a return trip to Miami from New York in the hot month of May (at least in south Florida) and not being very assertive as yet, it only took one woman passenger to flop her big fur coat over my arm. Before long I was up to my nose in fur coats, never realizing that I had to get all those furs tagged correctly and reunited with the correct owner in Miami. Not a good thing to give away an expensive fur coat to the wrong passenger! The proper procedure was to have the passenger take the coat with her to her seat and then I would come through and issue a seat check that would be half with her and the other half with the coat. Very simple. But you know, I never did figure out why anyone had to take a fur coat to Miami in the month of May!

"New York. New York. What a wonderful town" as the song says. There really was no greater thrill for me than to have a day "layover" in New York City. A whole day to see the entire city! Not possible, but so much fun trying to take in all the sights. The whole crew had been there before and knew a lot of ways to see the most in the least amount of time. I have been there several times since but nothing compared to that first trip. At night in New York, we "poor" stewardesses stayed in a very small hotel in Times Square. That was all we could afford in spite of receiving a small housing allowance from the company; by the end of the month we really had to watch our pennies. The "wealthy" captain and copilot stayed at fancier hotels in mid-Manhattan. My little room had a tiny balcony where I watched all the theaters empty out at about 11 p.m. The one that seemed to have the largest crowd was *No Time for Sergeants* with Andy Griffith. Broadway truly was "the great white way" in those days.

A typical inflight meal aboard a TWA 1649 Jetstream Starliner.

FROM BEST OF THE RADIALS TO THE FIRST JETS

The tail of a Boeing 377 Stratocruiser frames the new and then-revolutionary 707 jetliner at the dawn of the Jet Age in 1958. This is not only the end of piston engine airline transportation, but also of our book. Thank you for flying with us! (Photo Courtesy Boeing Co via Jon Proctor with permission)

As a passenger on one of Air France's Starliners emerges from the airplane and steps onto the ramp stairs, he is greeted by the loudest, harshest, highest-pitched noise he has ever heard in his life. Shocked at the howl, the traveler looks across the tarmac of New York's Idlewild Airport to see the source of the torment: one of those brand-new Boeing 707 jet airliners. *So,* this is the future, he thinks, and how right he is, perhaps more than he could imagine on this fall day in 1958. Thus, the page turns and reveals a new chapter and verse in the litany of aviation and human progress.

Barely 40 years after World War I, which introduced aerial combat to the notion of warfare and spawned a dynamic, authentically effective aeronautical technology and its application, here we were about to step through the looking glass with Alice and enter an entirely new and often foreign world. This was the turning point in global aviation which, when we looked back at it another several decades hence, signaled a complete shift in perspective toward our world, our neighbors, and ourselves.

Kicking and screaming, the airline business was being dragged into its inexorable future. There were many growing pains: infrastructure mostly, an all-too-frequent misunderstanding of the need for "speed-thinking," and the always-common human trait of wishing for exciting change but balking at its implementation. The Jet Age was dawning and was soon to overwhelm and smother the old world and its conception of getting there by air in the propeller days.

Things were happening fast, which did not help the transition. Air safety was as much a problem now with the jets as it had been in the DC-4 era but for many different reasons peculiar to the melding of the old mindset with the new. Technology was fresh out of the box and so was experiencing its normal (and sometimes abnormal) teething problems: Electras were shedding wings, 707s had a horrible Dutch Roll tendency, especially with two engines shut down on the same wing (several airplanes and crews were lost due to this problem), a few years prior to the Boeings the British Comet jetliner was disintegrating in flight due to fuselage metal fatigue and production shortcomings, and jets and propliners couldn't operate well together in the same airspace due to their tremendous speed differentials. People were dying with all this "new and improved" technology, and the sobering and contemporaneous question at that time was, "Surely this is going to get better?"

Yes. Human beings adapt and create, sometimes at unimaginable rates of speed. As this book is being written, the world is shedding one old skin for another at what seems like a daily pace. The people of the early 1960s thought the new jets made things incomprehensibly fast. They hadn't seen anything yet. But the world did begin thinking faster and, just as in the prior four decades, ever bigger as well. By 1964 it was supersonic transports (we were now getting used to jet speed!), then by 1969 it was Jumbo Jets. This is the testament to human ingenuity that is the basic theme and hallmark of international world-wide aviation and, therefore, this book.

If you are old enough, you can remember with fondness the lovely, halcyon days of luxurious, civilized airline travel and recount them with great joy. Times when flying was at first an epic adventure and then later a happy accomplishment on a routine basis, and finally, a magic carpet to the big, wide world.

Unfortunately, our memories always tend to preserve and present only the good parts and censor the less-than-savory aspects. Flying in those days was long and tedious, sometimes terrifying without the availability of weather radar or instrument landing systems. Vibration, noise, and a level of roughness always accompanied the journeys and, surprisingly, continued for hours to harmonize within every cell of our bodies after getting off the plane. Lavatories on airplanes were not much more than cans, repositories that did not really flush as we know them today. Some airplanes had decent air-conditioning systems on the ground, others kept you sweating the whole summer's day long until ascending to cooler altitudes.

And just as with the train and rail system in this country during the coal-burning days, flying was dirty. Radial engines needed oil, lots of oil. You know the old saying: When an engine is not dripping oil, that's when the pilot starts to worry. Well, inevitably that oil was on the concrete ramp; it was blown by the propellers all along the sides of the fuselage and over the wings, and it just seemed to be everywhere, on everything and everyone all the time. Once the airplane is bathed in this oily glaze, the next joy is all the dirt that adheres to it. What filth. And I would direct your attention back to the photo on the introduction page to review the oil-burning clouds that emanated from your average radial engine during all phases of its operation.

But you know, the period's vices were actually its virtues. Much of that gauzy, golden-hued aura really was a good thing. Life and airplanes were slower. The world had time to think, an act that used to be encouraged. There were fewer of us, both on the planet and sitting in airplane fuselages. Folks talked to one another then, a novel and quaint concept today. The world was a place to be visited and learned from, not just flown over and placed on a social media selfie page.

Above all, it was just plain fun to fly in a big, lumbering, shaking airliner. You really knew you were flying, unlike today when everything is lite or virtual or simulated. We smoked real, unfiltered cigarettes, drank honest whisky, and listened to actual music played by artisans on acoustic instruments. It was real men and real women using their creativity and gumption to get a big job done. Nothing was impossible; nothing was too far away. Those were the days of round engines and square deals.

Today it is ghosts. Everywhere we look, ghosts. Those evanescent wisps of reality that seem at once just minutes ago and once again, so far passed-by they must have been dreams. How fortunate we are to be able to recount the realities of days long gone, and express them now with photographic precision on glossy paper. For no matter how long ago any "past" may have been, it will always live vibrantly in our minds and hearts and inspire us to move ever forward into tomorrow's next yesterday.

Bibliography

Air Force and Bureau of Aeronautics. *Handbook Overhaul Instructions R-2000-4 Aircraft Engines*. 1 January 1950, Revised 1 May 1957.

Anderson, John D., Jr. *A History of Aerodynamics and Its Impact on Flying Machines*. Cambridge: Cambridge University Press, 1997.

Armbruster, G. E. "History of the Development of R-4360 Engines." *Service School Handbook*, unknown. http://www.enginehistory.org/Piston/P&W/R-4360/R-4360History.pdf. Accessed 13 October 2017.

Army Air Force. *Air Deport Progressive Overhaul Manual Pratt & Whitney R-1340-AN1 Engine*. 10 December 1943.

Avco Corporation. *Avco Corporation: The First Fifty Years*. Greenwich, Connecticut: Avco Corporation, 1979.

Bendix. *Bendix Electronic Ignition Analyzer*. Sydney, New York: Scintilla Magneto Division Bendix Aviation Corporation, February 1951.

Bingham, Victor. *Major Piston Engines of World War II*. Ramsbury, UK: Airlife, 1998.

Bowers, Peter M. *Boeing Aircraft since 1916*. Annapolis, Maryland: Naval Institute Press, 1989.

Breihan, John R., Stan Piet, and Roger S. Mason. *Martin Aircraft 1909–1960*. Santa Ana, California: Narkiewicz/Thompson, 1995.

Cameron, Kevin. "ATDC: Fixing the Wright R-3350 Radial." *Torque Meter Volume 7, Number 4*. Aircraft Engine Historical Society, Fall 2008.

———. "ATDC: R-3350 BA Reduction Gear Problems." *Torque Meter Volume 7, Number 1*. Aircraft Engine Historical Society, Winter 2008.

———. "ATDC: Why No B-29 Water-Injection?" *Torque Meter Volume 6, Number 2*. Aircraft Engine Historical Society, Spring 2007.

Cearley, George W. *American Airlines: An Illustrated History*. Self-Published, 1981.

———. *Western Air Lines: America's Oldest Airline*. Self-Published, 1987, 1991.

Connors, Jack. *The Engines of Pratt & Whitney: A Technical History*. Reston, Virginia: American Institute of Aeronautics and Astronautics, 2009.

Davies, R.E.G. *Pan Am: An Airline and Its Aircraft*. London: Orion Books, 1987.

———. *TWA: An Airline and its Aircraft*. McLean, Virginia: Paladwr Press, 2001.

Dicky, Philip S. III. *The Liberty Engine 1918–1942*. Washington, DC: Smithsonian Institute Press, 1968.

Dover, Ed. *The Long Way Home*. Self-Published by Ed Dover, 2010.

Eltscher, Louis R., and Edward M. Young. *Curtiss-Wright: Greatness and Decline*. Woodbridge, Connecticut: Twayne Publishers, 1998.

Fey, Tom. "Turbocompounding the Wright Way." *Torque Meter Volume 5, Number 2*. Aircraft Engine Historical Society, Spring 2006.

Francillon, René J. *Douglas Propliners: Skyleaders, DC-1 to DC-7*. Newbury Park, California: Haynes Publishing, 2012.

———. *Lockheed Aircraft Since 1913*. Annapolis, Maryland: Naval Institute Press, 1987.

———. *McDonnell Douglas Aircraft Since 1920: Volume I*. New York: Putnam, 1988.

Gann, Harry. *Douglas DC-6 and DC-7, Airliner Tech Volume 4*. North Branch, Minnesota: Specialty Press, 1999.

Ginter, Steve. *Martin Mars XPB2M-1R & JRM Flying Boats*. Simi Valley, California: Ginter Books, 1995.

Heron, S. D. *History of the Aircraft Piston Engine: A Brief Outline*. Richmond, Virginia: Ethyl Corporation, 1961.

Kholos, Len. *Courage and Innovation: The Story of LORD Corporation, 1924 to 2002*. Cary, North Carolina: Jura Corporation, 2005.

Klaas, M. D. *Last of the Flying Clippers: The Boeing B-314 Story*. Atglen, Pennsylvania: Schiffer Publishing Ltd., 1997.

Kuhns, Carl. "Turbocompounds." *Aircraft Engine Historical Society*, undated. http://www.enginehistory.org/Piston/Wright/Kuhns/CurtissWrightTC18/TurboCompounds.shtml. Accessed 5 October 2017.

Lage, Manuel. *Hispano Suiza in Aeronautics*. Warrendale, Pennsylvania: SAE International, 2004.

Larkins, William T. *The Ford Tri-Motor*. Atglen, Pennsylvania: Schiffer Publishing Ltd., 1992.

Lilley, Tom, et. al. *Problems of Accelerating Aircraft Production During World War II*. Cambridge, Massachusetts: Harvard University, 1947.

Lycoming. *Operators Manual: Lycoming R-680 and R-530 Aviation Engines, Second Edition*. Williamsport, Pennsylvania: Lycoming, June 1939.

McCutcheon, Kimble D. "Frank Walker: What can I do about this problem." *Aircraft Engine Historical Society*, undated. www.enginehistory.org/Biography/FrankWalkerWeb1.pdf. Accessed 16 October 2017.

———. "No Short Days: The Struggle to Develop the R-2800 "Double Wasp" Crankshaft." *Aircraft Engine Historical Society*, undated. http://www.enginehistory.org/engines.shtml. Accessed 16 October 2017.

———. *Tornado: Wright Aero's Last Liquid-Cooled Piston Engine*. Huntsville, Alabama: Weak Force Press, 2001.

———. "Torquemeters: Developments Through 1945." *Torque Meter Volume 4, Number 2*. Aircraft Engine Historical Society, Spring 2005.

———. "Wright J-5 Whirlwind." *Aircraft Engine Historical Society*, undated. http://www.enginehistory.org/Piston/Wright/WrightJ-5.pdf. Accessed 2 September 2017.

———. "Wright R-1820 Cyclone." *Aircraft Engine Historical Society*, undated. http://www.enginehistory.org/Piston/Wright/WrightR-1820.pdf. Accessed 2 September 2017.

———. "Wright R-3350 Cyclone 18." *Aircraft Engine Historical Society*, 27 October 2014. www.enginehistory.org/Piston/Wright/WrightR-3350.pdf. Accessed 10 October 2017.

Mead, Cary Hoge. *Wings Over the World: The Life of George Jackson Mead*. Wauwatosa, Wisconsin: Swannet Press, 1971.

Mitchell, Kent A. *Fairchild Aircraft 1926–1987*. Santa Ana, California: Narkiewicz/Thompson, 1997.

Pearce, William. "Republic XP-47J Superbolt Fighter." *Old*

Machine Press, 17 December 2013. http://oldmachinepress. com/2013/12/17/republic-xp-47j-superbolt/. Accessed 17 October 2017.

———. "Wright Aeronautical R-4090 Cyclone 22 Aircraft Engine." *Old Machine Press*, 22 March 2013. http://oldmachinepress. com/2013/03/22/wright-aeronautical-r-4090-cyclone-22. Accessed 11 October 2017.

Pratt & Whitney Aircraft Company. *Explanation of Pratt & Whitney Aircraft Engine Designations*. Pratt & Whitney, undated. http://www. enginehistory.org/Piston/P&W/PWdesignations.pdf. Accessed 22 September 2017.

———. *The Pratt & Whitney Aircraft Story*. Hartford, Connecticut: Pratt & Whitney, 1950.

———. *Pratt and Whitney Engine Handbook*. Hartford, Connecticut:Pratt & Whitney, 1929.

Pyeatt, Doan, and Dennis R. Jenkins. *Cold War Peacemaker: The Story of Cowtown and the Convair B-36*. Forest Lake, Minnesota: Specialty Press, 2010.

Raymond, Robert J. "Crankpin Bearings in High Output Aircraft Piston Engines." *Aircraft Engine Historical Society*, July 2015. www.enginehistory.org/members/articles/CrankpinBearings.pdf, Accessed 20 September 2017.

Rosen, George. *Thrusting Forward: A History of the Propeller*. Windsor Locks, Connecticut: United Technologies Corporation, 1984.

Serling, Robert J., and George H. Foster. *Steel Rails and Silver Wings*. Long Island, New York: Weekend Chief Publishing, 2006.

Scheppler, Robert H. "The Fokker F.32." *AAHS Journal. Vol. 11 No. 2*. American Aviation Historical Society, Summer 1966.

Schlaifer, Robert, and S. D. Heron. *Development of Aircraft Engines and Aviation Fuels*. Cambridge, Massachusetts: Harvard Business School, 1950.

Smith, Jay P. "Case History of the R-2600 Engine Project." *Aircraft Engine Historical Society*, undated. http://www.enginehistory.org/ Piston/Wright/R-2600/R-2600CaseHx.shtml. Accessed 4 October 2017.

———. "Case History of R-2600 Engine Project Lockland Investigation." *Aircraft Engine Historical Society*, undated. http://www. enginehistory.org/Piston/Wright/R-2600/R-2600Lockland.shtml. Accessed 4 October 2017.

Stringfellow, Curtis K., and Peter M. Bowers. *Lockheed Constellation*. St. Paul, Minnesota: Motorbooks International, 1992.

Sullivan, Kenneth H., and Larry Milberry. *Power: The Pratt & Whitney Canada Story*. Toronto: CANAV Books, 1989.

Thompson, Jonathan. *Vultee Aircraft 1932–1947*. Santa Ana, California: Narkiewicz/Thompson, 1992.

Unknown. "Compound Interest." *Flight Magazine*, 27 July 1950.

Unknown. "Lycoming Engines." *Wikipedia*. https://en.wikipedia. org/wiki/Lycoming_Engines. Accessed 15 October 2017.

Unknown, "Lycoming History: Decades of Pioneering Spirit." *Lycoming*. https://www.lycoming.com/history. Accessed 7 October 2017.

Unknown. "Lycoming R-680-BA, Radial 9 Engine." *Smithsonian National Air and Space Museum*. https://airandspace.si.edu/ collection-objects/lycoming-r-680-ba-radial-9-engine. Accessed 7 October 2017.

Unknown. *Wright Engines 1903 to 1940*. www.enginehistory.org/Piston/Wright/WrightThrough1940.pdf. Accessed 2 September 2017.

Van Deventer, John H., Jr. "The Story of Wright Aero." *Air Transportation*, 22 December 1928.

Veronico, Nicholas A. *Boeing 377 Stratocruiser, Airliner Tech Volume 9*. North Branch, Minnesota: Specialty Press, 2001.

Veronico, Nicholas A., and William T. Larkins. *Convair Twins, Airliner Tech Volume 12*. North Branch, Minnesota: Specialty Press, 2005.

Wegg, John. *General Dynamics Aircraft and Their Predecessors*. New York: Putnam, 1990.

White, Graham. *Allied Piston Aircraft Engines of World War II*. Warrendale, Pennsylvania: SAE International, 1995.

———. *R-2800: Pratt & Whitney's Dependable Masterpiece*. Warrendale, Pennsylvania: SAE International, 2001.

———. *R-4360: Pratt & Whitney's Major Miracle*. North Branch, Minnesota: Specialty Press, 2006.

Wiegand, F. J., and W. R. Eichberg. "Development of the Turbo Compound Engine" *SAE Transactions Volume 62*, 1954.

Wilkinson, Paul H. *Aircraft Engines of the World 1941*. Washington, DC: Paul H. Wilkinson, 1941.

———. *Aircraft Engines of the World 1951*. Washington, DC: Paul H. Wilkinson, 1951.

Wright Aeronautical Corporation. *Engines Shipped (on an invoice basis) by Wright Aeronautical Division and Licensees From 1920 to January 1, 1964*. Wright Aeronautical Corporation, September 1960, Revised June 1964. http://www.enginehistory.org/Piston/ Wright/C-WShipments1920-1964.pdf. Accessed 2 September 2017.

———. *Enginology*. Paterson, New Jersey: Wright Aeronautical Corporation, 1945.

———. *Facts about the Wright Turbo-Compound*. Paterson, New Jersey: Wright Aeronautical Corporation, October 1956.

———. *Historical Engine Summary (Beginning 1930)*. Wright Aeronautical Corporation, undated. http://www.enginehistory.org/Piston/ Wright/C-WSpecsAfter1930.pdf. Accessed 2 September 2017.

———. *Wright Aeronautical Engines*. Paterson, New Jersey: Wright Aeronautical Corporation, 1929.

———. *The Wright Cyclones*. Paterson, New Jersey: Wright Aeronautical Corporation, 1942.

———. *Wright Turbo-Compound*. Paterson, New Jersey: Wright Aeronautical Corporation, November 1952.

Wright Aeronautical Division. *Basic Theory of Operation Turbo Compound Engine*. Davidson, North Carolina: Curtiss-Wright Corporation, March 1965.

———. *Summary: Engines Shipped by Wright Aeronautical Division and Licenses*. Davidson, North Carolina: Curtiss-Wright Corporation, June 1964.

———. *Trouble Shooting for Optimum Performance*. Davidson, North Carolina: Curtiss-Wright Corporation, 1957.

Wolf, William. *Consolidated B-32 Dominator*. Atglen, Pennsylvania: Schiffer Publishing, 2006.

Index

Air Commerce Act of 1926, 13
Air France, 99, 102, 204, 205, 211
Air Mail Service, 12
American Airlines, 2, 6, 52, 55, 57–60, 63, 64, 67, 99, 104, 106, 120, 121, 139, 169, 176, 179, 180, 183, 213
American Overseas Airlines (AOA), 106, 120, 121
America's Round-Engine Airliners, 10, 202
Anzani, Alessandro, 25
Armagnac, 135
Arnold, Henry (Hap), 4, 125, 130
Atlantic Aviation, 17, 172
Atterbury, William, 14, 15
Aviation Corporation (AVCO), 49, 213
Aviation Traders, 116
Bach, 45, 47
Balzer, Stephen, 25
Bearcat, 160, 162, 194
Bendix, 165, 166, 213
Boeing 377, 7, 9, 10, 31, 120–122, 126, 135, 166, 167, 211, 214
Boeing 707, 9, 120, 142, 177, 211
Boeing 787, 9, 38, 94, 111, 207
Boeing Airplane Company, 47
Boeing Air Transport, 23, 53
Braniff, 137, 144, 170, 171, 173
British Overseas Airways Corporation (BOAC), 202
Browne, Kenneth, 197
Brown, Walter Folger, 19, 51
Canton, Georges, 25
Chance Vought Aircraft Company, 47
Cold War, 136, 214
Colonial Airlines, 18, 19, 51, 142
Constellation, 2, 7, 8, 103, 105–107, 120, 121, 136, 138–143, 147, 150, 151, 165–167, 174, 176, 178, 179, 183–185, 187, 194, 198–201, 203, 205, 208, 210, 214
Cord, Errett Lobban, 48, 49, 51
Corsair, 135, 162
Curtiss Aeroplane, 39, 71
Curtiss-Wright Corporation, 38, 39, 71, 206, 214
Daub, Rudolph (Rudy), 109, 112, 144, 151
Deeds, Edward, 45
de Havilland Canada, 116
Destroyer, 151
Dominator, 146, 151, 214
Douglas, Donald, 5, 7, 12, 47, 54, 55, 58, 105, 125, 179
Drinkwater, Terry, 178
Dynafocal, 197, 198

Edgar, Graham, 30
Electra, 7, 38, 68, 69, 75, 120, 177
Ellehammer, Jacob, 25
Federal Aviation Agency (FAA), 13, 76
Fleet, Rueben, 80
Flying Boxcar, 135, 194
Fokker, 7, 13, 17, 19, 20, 22, 38, 45, 47, 53, 99, 168, 214
Fokker, Anthony, 13
Ford, Henry, 13
Ford Motor Company, 13, 15, 17, 18, 38, 42, 45, 52, 53, 96, 106, 112, 135, 155, 161, 185, 195, 205, 213
Ford, Robert, 112
Frye, Jack, 7, 53, 55, 56
Globemaster, 135
Granger, Gordon, 209
Greenamyer, Darryl, 160
Guggenheim, Daniel, 14
Hall, Elbert J., 27
Hall-Scott Motor Car Company, 27
Hamilton Aircraft Company, 47
Hanshue, Harris, 16
Henderson, Paul, 14
Heron, Sam, 29, 33, 40, 41, 71
Hobbs, Leonard (Luke), 88, 125, 130, 152, 155
Hoover, Herbert, 13
Hornet, 21–23, 35, 43, 46, 47, 69, 71, 72, 79, 88, 107, 108, 207
Hughes Aircraft Company, 105, 117, 203, 208
Imperial Airways, 95
Japan Airlines, 202
Jones, Edward T., 40, 41
Kelly Act, 13
Keys, Clement, 14
KLM, 96, 104, 120, 183, 185, 202
Lawrance Aero Engines, 38
Lawrance, Charles, 38, 39, 41, 42, 70
Liberty engines, 12, 26–28, 37, 38, 213
Lindbergh, Charles, 14–16, 34, 41, 55, 80, 85
Lockheed Aircraft Corporation, 2, 4, 7, 35, 38, 39, 42, 45, 47, 54, 68, 69, 75, 76, 93, 105, 114, 117, 120, 138–140, 142, 143, 147, 150, 151, 162, 165, 167–169, 173, 178, 179, 183, 184, 193, 194, 200–203, 205, 208, 209, 213, 214
Lord Corporation, 197, 198, 213
Lord, Hugh, 197
Lord, Thomas, 197
Manly, C. M., 25
Mars, 146, 151, 213
Mauler, 135

Mead, George, 39, 40, 43, 71, 125, 155
Mercator, 135
Miston, Gerard, 209
Nash-Kelvinator, 155, 161
National Advisory Committee on Aeronautics (NACA), 22, 35, 36, 38, 145, 146, 150, 195
National Aeronautics and Space Administration, 35, 38, 145, 146
National Air Transport (NAT), 14
Neptune, 151, 194
Northrop Grumman Corporation, 6, 55, 75, 89, 93, 111, 112, 135, 153, 156, 160, 162, 194
Northrop, Jack, 55
NYRBA airline, 80
Pacific Air transport, 13, 19
Packard Motor Car Company, 27
Pan American Airways (PAA), 68, 78
Parkins, Wright, 152
Patterson, William "Pat", 95
Peacemaker, 135, 214
Pennsylvania Railroad (PRR), 14, 15
Philadelphia Rapid Transit Service (PRT), 13, 14, 178, 179, 188–191, 193, 194, 198
Pratt, Perry, 152
Pratt & Whitney Aircraft (P&W), 6, 9–11, 22, 23, 28–31, 33–35, 39, 41, 43–47, 53, 66, 71, 72, 75–77, 79, 88–93, 107–113, 116, 118, 119, 125, 130, 133–135, 146, 151–156, 158–164, 166, 167, 194–197, 205–207, 213, 214, 216
Qantas, 143, 183
Rainbow, 104, 117–120, 122, 135, 139
Raymond, Art, 54, 55
Rentschler, Frederick, 39, 41, 43, 71, 125, 155
Republic Aviation Corporation, 6, 93, 104, 117–120, 123, 135, 156, 162, 213
Ryder, Earle, 43
Salmson Company, 25
Sea Fury, 194
Sikorsky Aircraft Corporation, 20, 21, 35, 39, 42, 45, 47, 75, 79–83, 87, 93, 162
Skyraider, 151
Smith, Bill Littlewood, 58
Smith, Cyrus Rowlett, 58, 60, 64, 179
Speedpak, 143
Sperry, 165
Standard Steel Propellers, 47
Starliners, 193, 211
Stearman Aircraft Corporation, 47

Stout, William, 13
Stratocruiser, 7, 31, 114, 119–122, 124–126, 135, 167, 202, 211, 214
Stratofreighter, 135
Studebaker, 75, 111, 195
Super Constellation, 2, 151, 167, 178, 179, 183, 185, 194, 210
Superfortress, 134, 135, 146, 151, 167
Super G Constellation, 2, 167, 183, 187, 205, 208
Taylor, Edward, 197
Thunderbolt, 156, 162
Townend, Hubert C., 35
Trans Canada, 183
Transcontinental Air Transport (TAT), 14–20
Transcontinental and Western (TWA), 2, 4, 7, 20, 50, 53, 55–57, 64, 77, 80, 96, 100–102, 105, 106, 120, 123, 136, 139, 141–143, 168, 171–173, 178, 179, 183–187, 202–205, 208–210, 213
Trippe, Juan, 78–80, 82, 85, 114, 118, 120, 121
United Aircraft and Transport Company, 47
United Air Lines, 19, 46, 53, 126, 138, 176, 188
Unne, Pierre, 25
US Postal Service, 12, 13, 82
Varney, Walter, 13
Ventura, 147, 162
Vincent, Jesse G., 27
Vultee, 23, 55, 68, 93, 104, 112, 133, 154, 169, 214
Walker, Frank, 158, 213
Weick, Fred, 35
Wendel, Fritz, 160
Western Air Express, 13, 15, 16, 19, 22
Willgoos, Andy, 43, 71, 88, 152
World War I, 11, 12, 24–28, 30, 37, 88, 212
World War II, 6, 7, 30, 31, 42, 44–46, 49, 64, 65, 73, 74, 85, 89, 92, 96, 98, 103, 109, 111, 112, 116, 117, 136, 145, 151, 155, 157, 158, 160–162, 167, 195, 205, 213, 214
Wright Aeronautical, 2, 5, 10–13, 23–25, 29, 30, 34–43, 45, 47, 49, 55, 59, 69–73, 75–77, 88, 89, 91, 107–112, 133, 135, 138–140, 142, 144–152, 156, 159, 163, 164, 166, 167, 178, 179, 182, 188–190, 193–198, 201, 205–207, 213, 214
Wright brothers, 11, 24, 36–38, 196
Wright-Martin Aircraft Company, 37

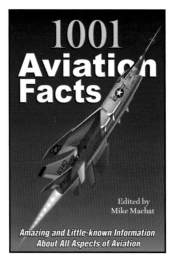
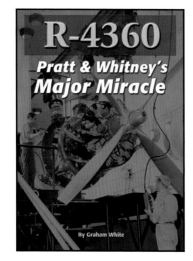
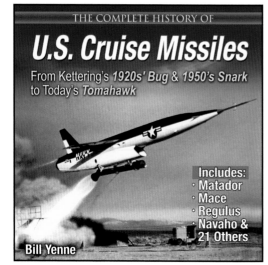
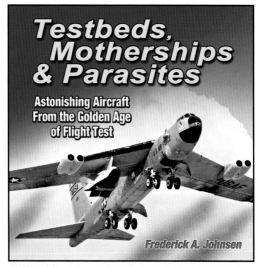

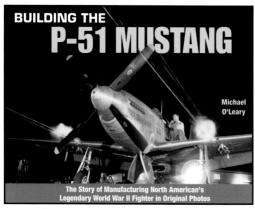